Deploying Secure Data Science Applications in the Cloud

From VMs to Serverless with AWS and Google Cloud

Lucas H. Benevides e Braga

Apress®

Deploying Secure Data Science Applications in the Cloud: From VMs to Serverless with AWS and Google Cloud

Lucas H. Benevides e Braga
Berlin, Germany

ISBN-13 (pbk): 979-8-8688-1714-4   ISBN-13 (electronic): 979-8-8688-1715-1
https://doi.org/10.1007/979-8-8688-1715-1

Managing Director, Apress Media LLC: Welmoed Spahr
Acquisitions Editor: Celestin Suresh John
Development Editor: Laura Berendson
Coordinating Editor: Gryffin Winkler

Cover image designed by Freepik (www.freepik.com)

Distributed to the book trade worldwide by Springer Science+Business Media New York, 1 New York Plaza, New York, NY 10004. Phone 1-800-SPRINGER, fax (201) 348-4505, e-mail orders-ny@springer-sbm.com, or visit www.springeronline.com. Apress Media, LLC is a Delaware LLC and the sole member (owner) is Springer Science + Business Media Finance Inc (SSBM Finance Inc). SSBM Finance Inc is a **Delaware** corporation.

For information on translations, please e-mail booktranslations@springernature.com; for reprint, paperback, or audio rights, please e-mail bookpermissions@springernature.com.

Apress titles may be purchased in bulk for academic, corporate, or promotional use. eBook versions and licenses are also available for most titles. For more information, reference our Print and eBook Bulk Sales web page at http://www.apress.com/bulk-sales.

Any source code or other supplementary material referenced by the author in this book can be found here: https://www.apress.com/gp/services/source-code.

If disposing of this product, please recycle the paper

Table of Contents

About the Author

Lucas Benevides e Braga is a Lead Data Scientist and AI Cloud Engineer with over ten years of international experience in data science. He has held senior and leadership roles like Staff and Manager Data Scientist positions at globally recognized organizations such as DHL, Delivery Hero, and DoorDash International (Wolt), where he led large-scale AI and cloud engineering initiatives.

Lucas holds a Master of Science in Applied Statistics and Data Science from Kansas University and has earned certifications and had multiple years of experience from both Google Cloud and Amazon Web Services. His expertise spans fraud detection systems, credit risk modeling, market intelligence, applied Large Language Models, and secure MLOps deployments.

With a deep commitment to knowledge transfer and technical enablement, Lucas has trained data teams, designed production infrastructure, and peer-reviewed systems, code bases, and IEEE research articles. He also authored data science and machine learning articles and was the recipient of multiple awards recognizing his leadership and technical expertise.

Recognizing a gap in the industry between secure deployment standards and practical guidance for data professionals, Lucas authored *Deploying Secure Data Science Applications in the Cloud* to help data scientists, ML engineers, and software engineers build secure, scalable, and production-ready systems. This book distills his frontline experience into actionable patterns and cloud-native practices for the next generation of applied AI builders. You can find him on LinkedIn at `www.linkedin.com/in/lucasbraga461/`.

About the Technical Reviewer

Neel Sendas is a distinguished professional in the field of cloud operations and machine learning, with nearly 20 years of experience. He has a proven track record of ensuring successful cloud operations and fostering strong client relationships.

Currently, Neel is working as a Principal Technical Account Manager at Amazon Web Services (AWS). In this role, he collaborates with cross-functional teams such as sales, architecture, and product management to prioritize customer feature requests, driving significant revenue growth for the business.

Prior to joining AWS, Neel was a senior consultant specializing in IoT and machine learning at Deloitte Digital, where he contributed to the development of innovative digital solutions. His career also includes a tenure as the vice president of digital strategy at Bank of America. Neel's early career was marked by his time as a software engineer at CA Technologies, where he soon advanced to the position of senior software engineer.

Academically, Neel holds a Master of Business Administration from the Tepper School of Business at Carnegie Mellon University. He also earned a Bachelor of Technology in Computer Science and Engineering from NE Hill University and completed additional coursework in accounting and finance at the Kenan-Flagler Business School at the University of North Carolina at Chapel Hill.

Acknowledgments

First, my gratitude to my wife Laura, to my parents, family, and to my friends. Your patience, encouragement, and willingness to share your time with a blinking cursor meant everything on the late nights and weekends when the manuscript took priority.

To my colleagues and managers, thank you for the many data science projects we developed together on AWS and Google Cloud. The lessons we learned became the pages you are now holding.

To Rodrigo, scientific research collaborator, DevOps Engineer wizard, and friend, thank you for that memorable session a few years ago that nudged me onto the MLOps Engineering path. That spark proved invaluable as I continued to build the expertise reflected in these pages.

I am indebted to my technical reviewer, whose careful reading and candid feedback kept the material clear, accurate, and practical.

Finally, thank you to Celestin for believing in the idea and giving it a home, and to Gryffin for steady guidance throughout the whole process. Your support turned a draft into a finished book that can now see the light of day.

Introduction

In today's landscape, for an application to be valuable, it needs to be secure; it must run behind proper access controls and encryption. For a company, an unprotected endpoint risks leaking customer data and invites automated attacks within minutes of going live. For you as an individual, hosting a side project on a laptop or an unprotected server leads to constant worry and manual supervision, and the very real possibility of waking up to a compromised machine.

Secure cloud deployment solves both problems: it wraps your code in authentication, encryption, and least-privilege access so the business avoids costly breaches, and it lets your ad hoc ideas operate with the polish and reliability of a production service, 24x7, no human supervision required.

Why This book, and Why Now?

Cloud platforms, container runtimes, CI/CD pipelines, and serverless services have matured to the point where a single practitioner can launch production-grade systems in hours rather than months. Yet the books and tutorials for these tools treat security as a footnote, if it mentions it at all. Organizations are therefore reluctant to expose a data science prototype, no matter how brilliant, when an unsecured endpoint could leak confidential data to the entire internet.

When secure-deployment guides do exist, they are written for DevOps specialists and rarely connect the dots end to end from a data science perspective. The result is a persistent skills gap: many data scientists can train an outstanding model but feel uncertain, sometimes intimidated, when asked to deliver it as a secure, scalable service.

This book closes that gap. Project by project, you will take familiar Python artifacts, a Jupyter notebook, a pickled model, and a Streamlit script and walk them all the way to secure deployment on both AWS and Google Cloud. Along the journey you will adopt the DevOps habits–version control, containerization, and infrastructure hardening– that turn ad hoc notebooks, dashboards, and ETL jobs into production systems your company can trust.

Who Should Read This Book?

This book is written for **beginner-to-intermediate data professionals** who code in Python, feel comfortable with basic Linux commands and cloud terminology, yet have little hands-on experience deploying their work. Whether you are a data scientist, ML engineer, data engineer, or a DevOps/software engineer curious about data science workloads, each chapter walks in small, reproducible steps and uses language you already know. No prior expertise in Docker, Nginx, or network security is assumed; you will build those skills incrementally as the projects unfold.

What You Will Learn

- Deploy end-to-end data science applications with a strong foundation in cloud infrastructure setup, including VM provisioning, SSH access, security groups, SSL configuration, load balancers, and domain management

- Use industry-known tools such as Docker, Nginx, Flask, Streamlit, and Jenkins to build secure, scalable services

- Structure and expose machine learning models via APIs for production use

- Explore modern serverless architectures with AWS Fargate and Google Cloud Run to scale efficiently with minimal overhead

- Develop a cloud deployment mindset grounded in doing things from scratch—before adopting abstracted solutions

How the Book Is Organized

The journey is split into three parts:

- **Part 1, Building the Foundation**, includes Chapters 1–6 that walk from a blank AWS account to a fully secured, load-balanced stack running Jenkins, Flask, and Streamlit behind custom subdomains. The same architecture is then reproduced, and improved, in Chapters 7 and 8 on Google Cloud.

- **Part 2, Serverless Deployments**, shows how to re-deploy those services on Google Cloud Run (Chapter 9) and AWS ECS Fargate (Chapter 10), trading servers you manage for infrastructure the provider manages on your behalf.

- **Part 3, Jenkins, Streamlit, and Flask Demos**, zooms into three concrete use cases: a Jenkins ETL job (Chapter 11), a Streamlit business dashboard fed by that ETL (Chapter 12), and a Flask microservice that serves a trained model through a /predict endpoint (Chapter 13).

Each chapter ends with a short "Summary" section so you can verify that every step ran as expected before moving on.

A Note on Security

Throughout the book security is treated as a first-class requirement, not an afterthought. You will tighten inbound rules, encrypt traffic end to end, restrict credentials to least privilege, and store secrets outside of code, because production systems, even hobby projects, deserve that respect.

How to Use This Book

Follow the chapters in order if you are starting from scratch; each builds on the previous one. If you are already comfortable with, say, Cloud Run or Streamlit, feel free to skip ahead; with prerequisites listed up front and all code available in the companion GitHub repository.

By the final page, you will have travelled the full path from notebook to production, armed with deployable templates you can adapt to your own organization. More importantly, you will have gained the confidence to design secure, scalable workflows whenever your next data science idea is ready to meet its users.

Welcome to *Deploying Secure Data Science Applications in the Cloud*. Let's make your data science applications see the light of day.

PART I

Building the Foundation

## CHAPTER 1

Initial Setup on Your AWS Account

To begin deploying secure and scalable applications on AWS (`aws.amazon.com`), you need to set up your account and environment properly. This chapter walks you through the initial steps, including creating an IAM user, launching an EC2 instance, setting up security groups, and configuring an Elastic IP. These foundational steps will prepare your environment for deploying data science applications later in the book.

Set Up IAM Users

If you've just created your AWS account, you're logged in as the root user. The first step is to enable Two-Factor Authentication (2FA) for the root account and create an IAM user with limited permissions to follow the principle of least privilege.

The root user has unrestricted access to all resources, which makes it uncommon that developers have root user access in corporations, and also it is unnecessary for most purposes including for the purpose of productionizing ML applications (the purpose of this book).

Steps to Create an IAM User

1. Log in to the AWS Management Console and navigate to **IAM ➤ Users**.

2. Click **Create user** and provide the following details:

   - **User name:** Use a descriptive name, such as `developer-user`.

   - **Access type:** Select **Provide user access to the AWS Management Console - *optional***, click **I want to create an IAM user**, and generate a password. This option is important because this developer user will need programmatic access later on.

3. Click **Next** and proceed to create a user group (see Figure 1-1):

    - Name the group (e.g., `developer-group`).

    - Assign necessary policies to the group, such as

        - `AmazonEC2FullAccess`

        - `AmazonS3FullAccess`

        - `ElasticLoadBalancingFullAccess`

4. Add the IAM user to this group and click **Create user**.

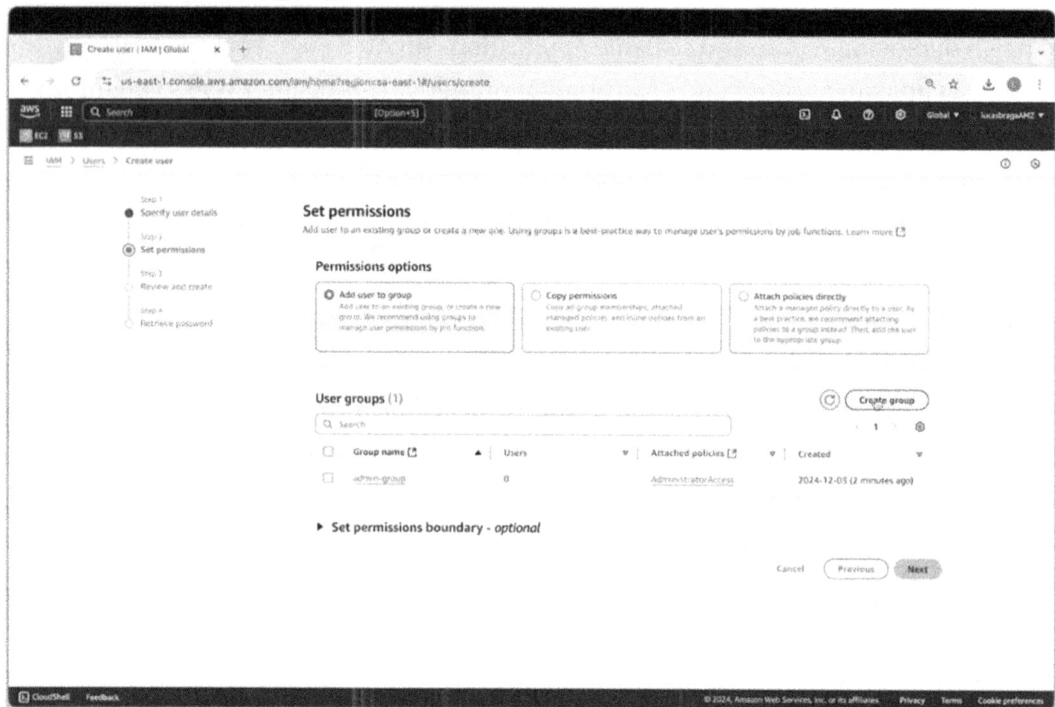

Figure 1-1. *IAM user creation page*

Once created, log out of the root account and log in with the new IAM user credentials. This ensures you're working in a more secure environment.

Create an EC2 Instance

An EC2 instance is a virtual server in AWS where you can host your applications. In this step, you'll launch an instance suitable for deploying multiple applications.

Steps to Launch an EC2 Instance

1. Navigate to **EC2 ➤ Instances**, then click **Launch instance**.

2. Configure the instance with the following settings:

   - **Name**: `my-web-server`

   - **AMI**: Select the default Amazon Linux 2 AMI.

   - **Instance type**: `t2.xlarge` (4 vCPUs, 16 GiB memory) is a good option to accommodate multiple applications like hosting Flask, Streamlit, and Jenkins simultaneously.

   See Figure 1-2.

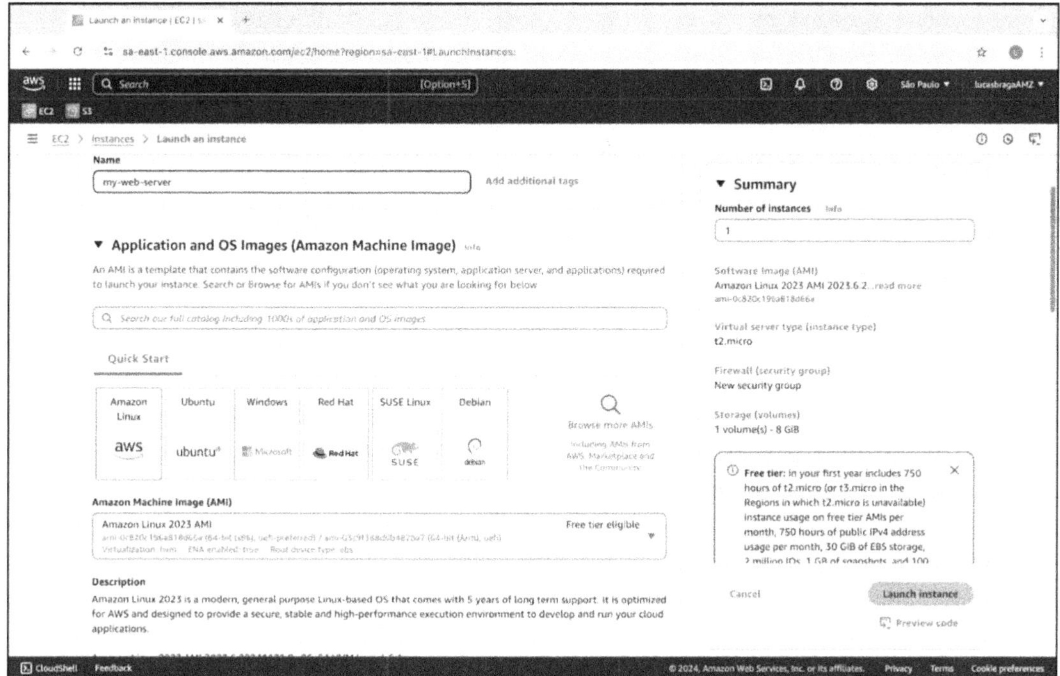

Figure 1-2. *Selecting the default Amazon Linux 2 AMI*

3. Create a key pair:

- Click **Create a new key pair**, name it (e.g., `my-web-server-key-pair` as shown in Figure 1-3), and download the `.pem` file. Store it securely.

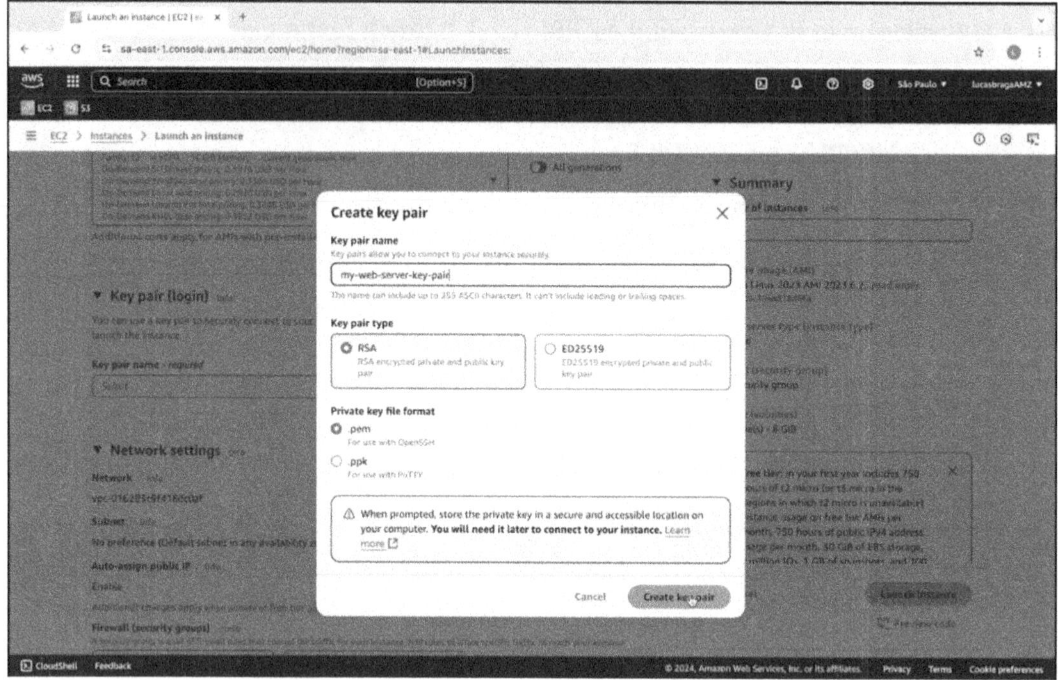

Figure 1-3. *Creating a key pair*

4. Adjust the storage:

- Set the root volume to at least 50 GiB to handle Docker image caching and temporary files. **100 GiB** might be better at the beginning because when running Docker up and down multiple times, the images are cached and this can consume a significant amount of memory.

5. Review the configuration and click **Launch instance**; see Figure 1-4.

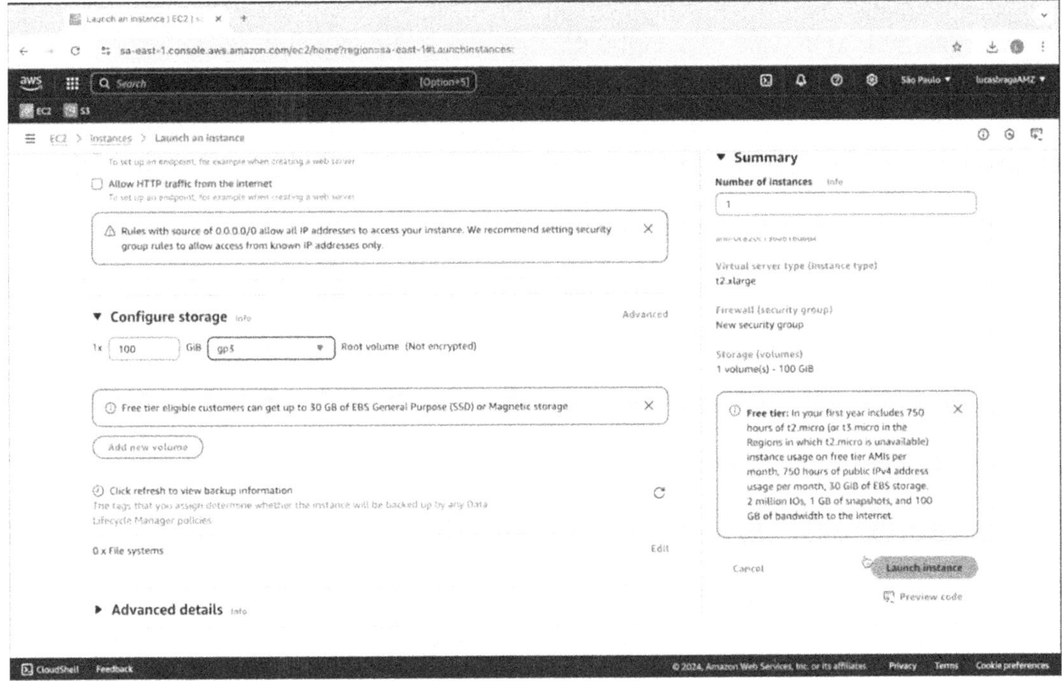

Figure 1-4. *Increasing storage to 100 GiB and launching instance*

Configure Security Groups

The first deployment in this book will demonstrate how to deploy an application directly on the EC2 instance. This is the simplest method of deployment and can be secure if configured correctly. Allowing access only to specific IP addresses ensures the safety of your application. However, combining this setup with a load balancer and an SSL certificate, as covered in Chapter 3, "Load Balancer on Your AWS Console," and Chapter 4, "Domain Name and SSL Certificates," provides even greater security.

To proceed, first identify the IP addresses or ranges that should have access to your instance. For example, to allow access only from your IP address, use a tool like `https://whatismyipaddress.com/` to find your IP. Append `/32` to specify a single IP address, for example, `35.123.123.58/32`. If you're connected to a VPN, the IP address shown will be often the VPN's.

If your organization uses a corporate VPN, you can grant access to the IP address behind that VPN so that only employees using the VPN have access to the application. Restricting access to trusted sources minimizes exposure to unauthorized traffic.

Note   Restricting access to specific IPs or VPNs minimizes the risk of unauthorized access.

A security group acts as a virtual firewall for your instance, controlling inbound and outbound traffic. Proper configuration is essential to secure your applications.

Steps to Create a Security Group

1. Navigate to **EC2 ➤ Security Groups**, then click **Create security group**.

2. Configure the following details:

   - **Name**: `ec2-security-group`

   - **Description**: "Allow SSH and application ports."

3. Add inbound rules, all with Protocol TCP and to the same IP address (e.g., `35.123.123.58/32`), as the following:

   - Type: **SSH (port 22)**. Optional description: "my ip - ssh access"

   - Type: **Custom TCP**, port **8501.** Optional description: "my ip - Streamlit access"

   - Type: **Custom TCP**, port **8502.** Optional description: "my ip - Flask access"

   - Type: **Custom TCP**, port **8504.** Optional description: "my ip - Jenkins access"

4. Leave the outbound rules as the default (allowing all traffic).

5. Attach this security group to your EC2 instance (**EC2 ➤ Instances**, then select your instance, and on the **Actions** button, click **Security ➤ Change security groups**; associate the instance with the security group that you've created and save it).

Set Up an Elastic IP

An Elastic IP provides a fixed public IP address for your instance, ensuring consistent access even after reboots. This is beneficial because, without an Elastic IP, the EC2 instance would receive a new public IP address each time it is restarted, requiring you to look up the new IP address every time you connect via SSH.

Steps to Allocate and Associate an Elastic IP

1. Navigate to **EC2 ➤ Elastic IPs** (under Network & Security).

2. Click **Allocate Elastic IP address**, then click **Allocate**.

3. Select the newly allocated Elastic IP (e.g., `54.94.167.229`) and click **Actions ➤ Associate Elastic IP address**.

4. Configure the association:

   - **Resource type**: Select `Instance`.

   - **Instance**: Choose your EC2 instance (e.g., `my-web-server`).

   - **Private IP**: Select the private IP of your instance (e.g., `172.31.8.127`; you can double-check that on EC2 ➤ Instances).

5. Click **Associate** and verify that the Elastic IP is now assigned to your instance.

Summary

You have now implemented the foundational layers for a secure deployment of your applications on an AWS EC2 instance; we have first

- Created an IAM user with the necessary permissions

  - IAM user documentation: `https://docs.aws.amazon.com/IAM/latest/UserGuide/id_users_create.html`

- Launched an EC2 instance with the appropriate size

  - EC2 getting started guide: `https://docs.aws.amazon.com/AWSEC2/latest/UserGuide/EC2_GetStarted.html`

- Configured security groups to restrict access to specific IP addresses or VPN users.

  - Security groups documentation: `https://docs.aws.amazon.com/AWSEC2/latest/UserGuide/ec2-security-groups.html`

- Assigned an Elastic IP to maintain consistent network access

  - Elastic IP documentation: `https://docs.aws.amazon.com/AWSEC2/latest/UserGuide/elastic-ip-addresses-eip.html`

Additional security configurations will be introduced in later chapters. Next, the following chapter will guide you through accessing your EC2 instance via SSH and configuring it for development.

## CHAPTER 2

SSH to the EC2 Instance with VS Code and Necessary Setup

In the previous chapter, you have successfully set up foundational AWS infrastructure by creating an IAM user, launching an EC2 instance, configuring security groups, and assigning an Elastic IP. In this chapter, you will continue building upon this foundation by learning how to securely access your EC2 instance through SSH, leveraging Visual Studio Code (VS Code) for better convenience. You'll also set up essential tools such as Docker, Docker Compose, GitHub, and Python with a virtual environment. Finally you'll deploy a simple HTTP application using Streamlit to validate your setup and ensure everything is functioning correctly. All code for this and every chapter is in the companion GitHub repo: `github.com/lucasbraga461/deploy-secure-ds-apps-book`. Some figures still show the old folder name my-web-server—that was the previous repo name. When you clone the repo you'll get a folder called deploy-secure-ds-apps-book (e.g., `~/Documents/GitHub/deploy-secure-ds-apps-book`).

SSH Connection to the Instance Using VS Code

One method to access the Linux terminal of the virtual machine (VM) created in AWS is through the AWS Management Console. Navigate to **EC2 ➤ Instances**, select your instance, and click **Connect**. On the SSH page, you will find instructions to execute a command similar to the following in your terminal; see Code-block 2-1.

Code-block 2-1. SSH connection

```
ssh -i /path_to_private_key/private_key.pem ec2-user@54.94.167.229
```


© Lucas H. Benevides e Braga 2025
L.H.B.e. Braga, *Deploying Secure Data Science Applications in the Cloud*,
https://doi.org/10.1007/979-8-8688-1715-1_2

Here's a breakdown of the command from Code-block 2-1:

- **/path_to_private_key/private_key.pem**: Replace this with the path to the private key you created in Chapter 1 during the instance setup.

- **ec2-user**: This is the default username for Amazon Linux AMI instances.

- **54.94.167.229**: Replace this with the public IP address of your EC2 instance. You can locate the IP in the AWS console by navigating to **EC2 ➤ Instances** and selecting your instance.

While this method provides direct access to your instance, it is not the most efficient or user-friendly approach. In the section "SSH Connection to the Instance Using VS Code," we will demonstrate how to establish an SSH connection using **VS Code**, which offers an integrated and streamlined experience for managing your instance.

Connecting to your EC2 instance using VS Code provides more than just terminal access. VS Code allows you to interact with your instance's file system through an integrated file explorer and use text editors with syntax highlighting for various programming languages.

Steps to Connect to the Instance

1. **Prepare the SSH configuration**

   - On your local machine, create a `.ssh` folder in your home directory if it doesn't already exist; see Code-block 2-2.

   Code-block 2-2. Create a folder using Linux Bash language

   ```
   mkdir ~/.ssh
   ```

   Inside the `.ssh` folder, create a file named `config` and add the following configuration, as shown in Code-block 2-3.

   Code-block 2-3. .ssh/config file

   ```
   Host lb-ec2-mywebserver
       HostName 18.228.131.1
       IdentityFile ~/Documents/config/my-web-server-key-pair.pem
       User ec2-user
   ```

Replace 18.228.131.1 with your instance's public IP, which you can find in the AWS Management Console under **EC2 ➤ Instances**.

2. **Install the VS Code Remote Explorer Extension**

   - Open VS Code, navigate to the Extensions Marketplace (left panel), and install the **Remote - SSH** extension by Microsoft.

   - After installation, you'll see on the left panel a new symbol (Remote Explorer) that looks like a TV.

3. **Establish the SSH connection**

   - In VS Code, click the **Remote Explorer** icon on the left panel.

   - Refresh the view, and you should see the entry lb-ec2-mywebserver listed. Click **Connect in New Window** to initiate the SSH connection.

Note   Ensure your security group allows SSH (port 22) access from your IP address or VPN.

Install Docker and Docker Compose

Docker is a powerful tool for creating standardized, reproducible environments for your applications. Docker Compose extends this capability by orchestrating multiple containers. Let's begin by installing Docker and Docker Compose on your EC2 instance.

Steps to Install Docker

1. **Install Docker**

   Run the following commands in the terminal, shown in Code-block 2-4.

   Code-block 2-4. Installing Docker

```
sudo yum install docker -y
docker --version
```

Note that yum is the command used to do installations in this Linux version (the default Linux option from AWS that you selected). However, if you have selected another Linux version and this command is not working, replace yum by apt-get. The option -y at the end of the command is to say yes to all possible confirmation questions that come with the installation.

The second command docker --version is to verify that Docker was installed; it returns the Docker version; see Figure 2-1.

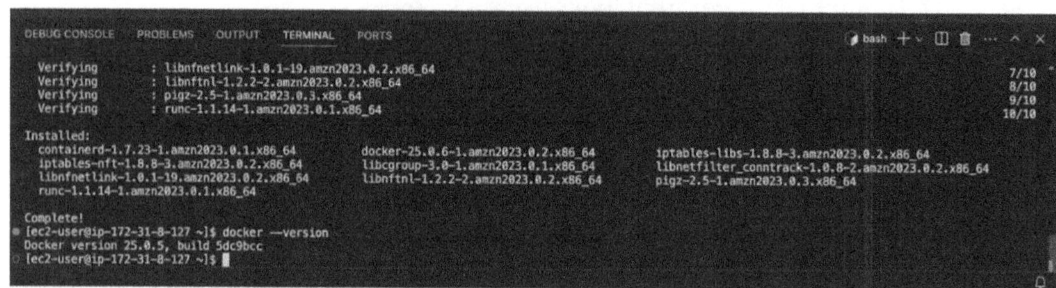

Figure 2-1. *Docker version from the terminal in VS Code*

2. **Add Docker's official repository and install docker-ce**

    Run the following commands, from Code-block 2-5, in the terminal.

    Code-block 2-5. Adding Docker's official repository

```
sudo yum-config-manager --add-repo https://download.docker.com/
linux/centos/docker-ce.repo
sudo yum upgrade
sudo yum install docker-ce -y
```

3. **Enable and start Docker**

    After deploying the application with Docker, if you shut down the EC2 instance and then restart it, the application will not automatically run. You would need to reconnect via SSH and manually restart Docker to get the application running again. To avoid this, you can enable the Docker service to start automatically when the instance is powered on. This is particularly useful in scenarios where the EC2 instance is scheduled to shut down during non-working hours and restart during working hours. By enabling the service, you eliminate the need to SSH into the instance every time, ensuring the application starts up seamlessly. Refer to the commands in Code-block 2-6 to configure this behavior.

Run the following commands, from Code-block 2-6, in the terminal.

Code-block 2-6. Enable Docker service in the instance

```
sudo systemctl start docker
sudo systemctl enable docker
sudo systemctl status docker
sudo docker info
```

The commands in Code-block 2-6 serve distinct purposes:

- `start`: This starts the Docker service, making it ready for immediate use. However, this alone does not configure the service to start automatically if the instance is rebooted.

- `enable`: This command ensures the service is configured to restart automatically when the instance reboots.

- `status`: Use this to check the current status of the Docker service.

- `docker-info`: Displays system-wide information about the Docker installation.

Figure 2-2 illustrates the Docker status on the instance when the service has not been enabled to start automatically. Note that the service is **active** (highlighted in green), but its automatic startup is **disabled** (highlighted in yellow).

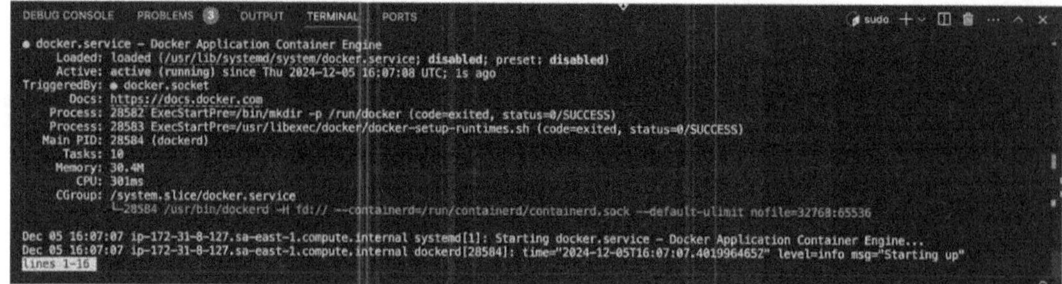

Figure 2-2. *Docker status before running sudo systemctl enable docker*

Steps to Install Docker Compose

1. **Set up the folder and install docker-compose**

   Breaking down the commands from Code-block 2-7:

   - DOCKER_CONFIG=: That is a variable with the location of the folder that will be used for docker-compose installation.

   - mkdir: This command creates a folder, and it uses the variable DOCKER_CONFIG for that.

   - curl -SL: This command downloads docker-compose from the internet.

   - chmod +x: This command gives execution rights to allow docker-compose to be used.

   Code-block 2-7. docker-compose installation

```
DOCKER_CONFIG=${DOCKER_CONFIG:-$HOME/.docker}
mkdir -p $DOCKER_CONFIG/cli-plugins
curl -SL https://github.com/docker/compose/releases/download/
v2.6.1/docker-compose-linux-x86_64 -o $DOCKER_CONFIG/
cli-plugins/docker-compose
chmod +x $DOCKER_CONFIG/cli-plugins/docker-compose
```

2. **Add Docker Compose to the path**

   Update the .bashrc (usually located at the home directory: /home/ec2-user/) file to include Docker Compose; see Code-block 2-8 and Figure 2-3.

Code-block 2-8. Add these lines to the .bashrc file

```
alias docker-compose='/home/ec2-user/.docker/cli-plugins/
docker-compose'
export PATH="/home/ec2-user/.docker/cli-plugins/:$PATH"
```

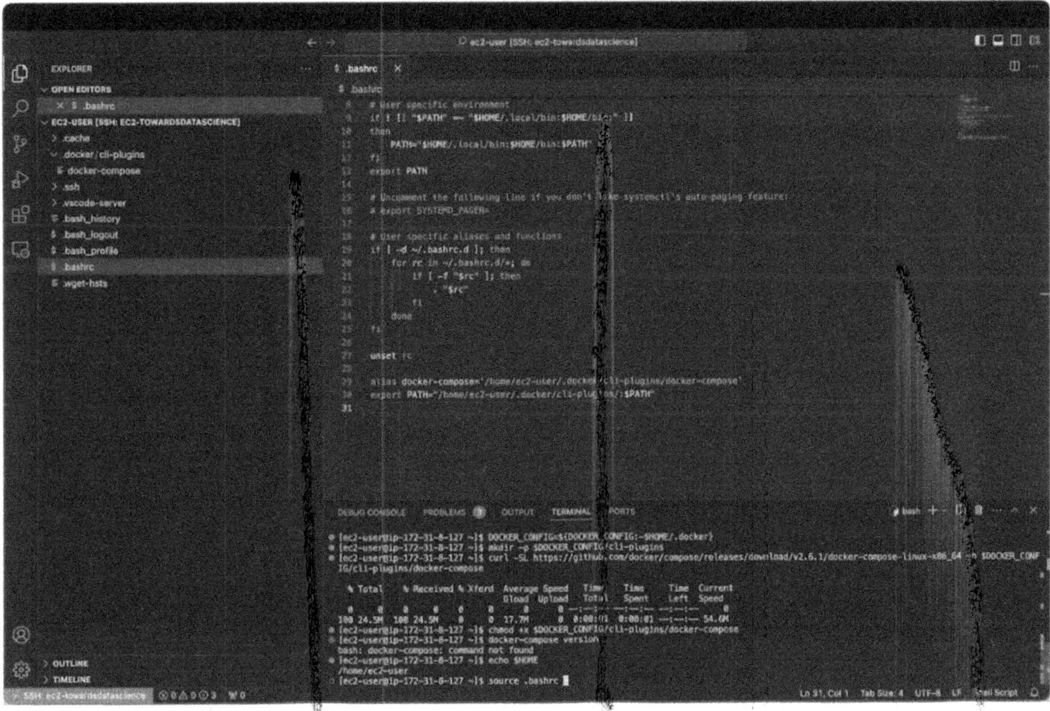

Figure 2-3. *Source .bashrc*

Now run the following command from Code-block 2-9 to execute
the changes and to link the local path to a path that the root user
can access so that when you run the command with sudo, it
will work.

Code-block 2-9. Run this command on the terminal to execute
the changes

```
source ~/.bashrc
sudo ln -s /home/ec2-user/.docker/cli-plugins/docker-compose /usr/
bin/docker-compose
```

3. **Test the installation**

- Verify that Docker Compose is installed correctly; see
  Code-block 2-10 and Figure 2-4.

Code-block 2-10. Verify docker-compose installation

```
docker-compose --version
```

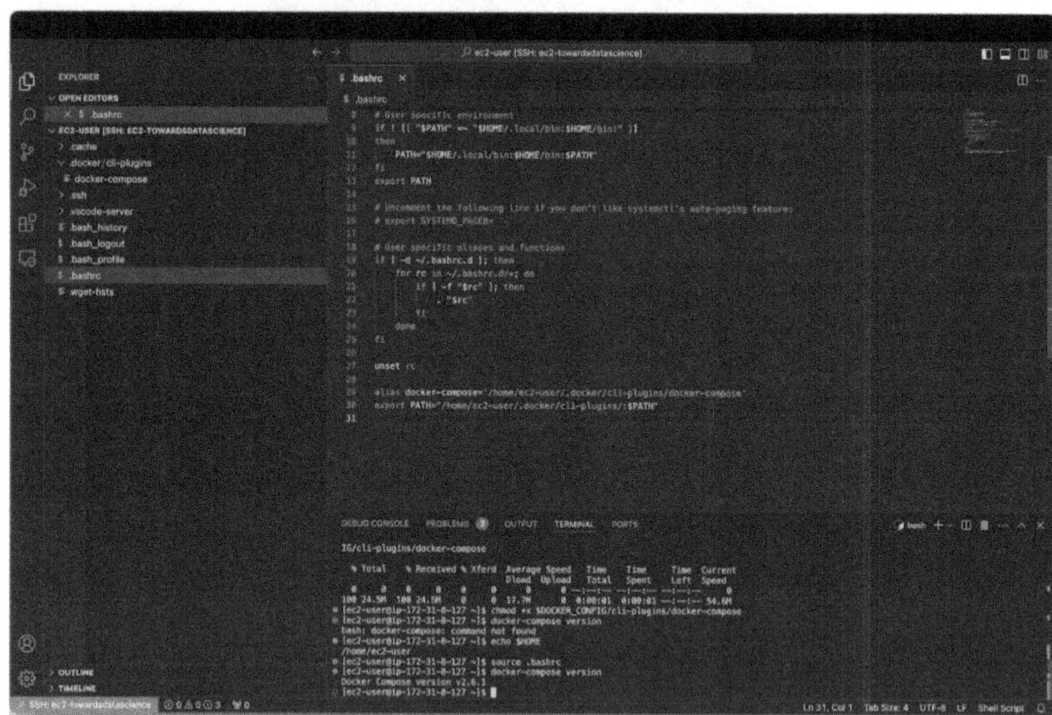

Figure 2-4. *Installing and configuring Docker and Docker Compose*

Set Up GitHub and Clone the Repository

GitHub is essential for managing and version-controlling your application's code. This
section guides you through setting up GitHub access, configuring SSH keys, and cloning
a repository to your EC2 instance.

Steps to Configure GitHub Access

1. **Generate SSH keys**

   Run the following commands, from Code-block 2-11, to create
   SSH keys:

   Code-block 2-11. Create public and private SSH keys

   ```
   ssh-keygen -t ed25519 -C "your_email@gmail.com"
   eval "$(ssh-agent -s)"
   ```

   - This generates a public key (id_ed25519.pub) and a private key
     (id_ed25519) in the ~/.ssh folder. Remember to replace "your_
     email@gmail.com" by the actual email associated with your
     GitHub account.

2. **Look for a ~/.ssh/config file**

   - Within the ~/.ssh folder, check also if there is a ~/.ssh/config
     file; if there isn't, create one and add the following according to
     Code-block 2-12.

     Code-block 2-12. .ssh/config file

     ```
     Host *.github.com
         AddKeysToAgent yes
         IdentityFile ~/.ssh/id_ed25519
     ```

3. **Add the private key to SSH agent**

   - Ensure your private key is loaded into the SSH agent; see Code-
     block 2-13.

     Code-block 2-13. Run this on the terminal if you didn't have the
     .ssh/config

     ```
     ssh-add ~/.ssh/id_ed25519
     ```

4. **Configure GitHub SSH access**

- Copy the contents of your public key. You can either manually in VS Code click and open the file that contains the public key `~/.ssh/id_ed25519.pub` and copy it, or run the following command, from Code-block 2-14, on the Linux terminal.

Code-block 2-14. Copy the public key

```
cat ~/.ssh/id_ed25519.pub
```

- Go to your GitHub account, navigate to **Settings ➤ SSH and GPG Keys**, and add a new SSH key. Give it a descriptive title like `ec2-mywebserver`, and then we paste the public key there under Key.

- Once done, click `Add SSH key` to add it.

    i. **If it's a corporate GitHub account**, an additional security step might be required. Check on the right of the newly added SSH key whether you have a button that says `Configure SSO`; if so, click the drop-down arrow and click `Authorize` the organization. See Figure 2-5 to confirm how it should look after you authorized the organization with SSO.

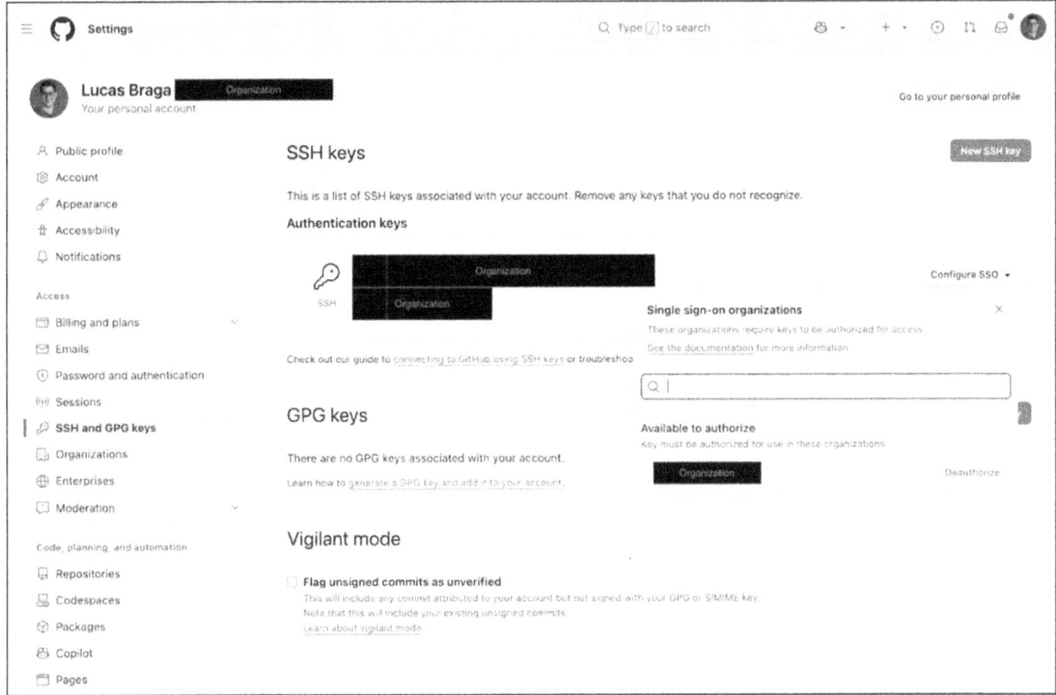

Figure 2-5. *SSO authorized*

5. **Test the connection**

   - Back to the VS Code terminal, test whether the connection was successfully established with the following commands from Code-block 2-15. Note that a question usually comes in as `Are you sure you want to continue connecting (yes/no/ [fingerprint])?` Then reply yes on the terminal.

   Code-block 2-15. Test the SSH connection on the terminal

   ```
   ssh -T git@github.com
   yes
   eval "$(ssh-agent -s)"
   ```

6. **Secure the GitHub keys**

   - Secure the public and private GitHub keys with the commands from Code-block 2-16.

Code-block 2-16. Secure the public and private keys

```
chmod 700 .ssh/
chmod 400 .ssh/config
```

Understanding the chmod command from the Linux Bash language:

- The chmod command in Linux is used to modify file and folder permissions. Permissions are defined using a combination of numbers that represent basic actions:

  i.   **Read (r)**: Value of 4

  ii. **Write (w)**: Value of 2

  iii. **Execute (x)**: Value of 1

- Permissions for a user, group, and others are represented as a sequence of three numbers. Each number is the sum of the values corresponding to the granted permissions. For example:

  i.   A user granted both read and write permissions would have a value of 6 (4 + 2).

  ii. The sequence 700 means

       - **User**: Has read, write, and execute permissions (7 = 4 + 2 + 1)

       - **Group**: Has no permissions (0)

       - **Others**: Has no permissions (0)

- This system provides a simple yet powerful way to control access to files and directories.

Steps to Clone the Repository

1. **Install git (if necessary)**

   - Double-check whether you have git already installed by running the following code on the terminal; see Code-block 2-17. In case it returns a git version, the installation is already done. In case it returns command not found, install it; see Code-block 2-18.

Code-block 2-17. Check if git is installed

```
git version
```

Code-block 2-18. Install git

```
sudo yum install git -y
```

2. **Clone the repository**

Navigate to your desired folder and clone the repository; see Code-block 2-19.

- Note: To find the correct path for use with the git clone command, navigate to the GitHub repository's web page, click the <> Code button, select the SSH option, and copy the displayed URL.

Code-block 2-19. Clone the repository

```
mkdir -p ~/Documents/GitHub
cd ~/Documents/GitHub
git clone git@github.com:lucasbraga461/deploy-secure-
ds-apps-book.git
```

Finally, create a .gitignore file within the repository's main folder to specify files that should not be committed to the main branch (see this file in the GitHub repository).

Next, set up a secondary branch, referred to here as first_commits, for development purposes. This ensures that changes are not made directly to the main branch, which is good practice in code development. Once development on the secondary branch is complete, you can create a pull request to merge the changes into the main branch. Code-block 2-20 shows how to create a secondary branch.

Code-block 2-20. Clone the repository

```
cd ~/Documents/GitHub/deploy-secure-ds-apps-book
git checkout -b first_commits
```

Install Python and Set Up a Virtual Environment

Python is essential for running data science applications, it provides the foundation for many libraries and frameworks. Setting up Python with a virtual environment is key to isolate dependencies and keep them manageable, which prevents conflicts between projects. Virtual environments create a clean workspace for each project, allowing you to install only the packages you need without affecting your global Python setup.

If you work in a single environment without using virtual environments, you risk dependency issues as package versions conflict across projects. While installing packages globally may seem simpler for a single project, it lacks the flexibility and scalability needed as your workflows grow. Using virtual environments is a best practice for maintaining a clean and conflict-free setup, especially in complex or multi-project environments.

Steps to Install Python

1. **Download and install Python**

   To install Python on your EC2 instance, start by accessing the terminal and navigating to the home folder with `cd ~` and download the Python package with `sudo wget`. Once downloaded, unpack the package with `sudo tar xzf` and navigate into the unpacked Python folder.

   From there, run the initial installation commands and proceed to install the `make` command. Afterward, delete the `.tgz` package you downloaded since it's no longer needed after unpacking. Finally, complete the installation by running the `make altinstall` command. With these steps, the Python installation will be complete. Run the following commands from Code-block 2-21.

   Code-block 2-21. Install Python

```
cd ~

sudo wget https://www.python.org/ftp/python/3.9.9/Python-3.9.9.tgz

sudo tar xzf Python-3.9.9.tgz

cd Python-3.9.9
```

```
sudo ./configure --enable-optimizations

sudo yum install make

sudo rm -f /Python-3.9.9.tgz

sudo make altinstall

python3.9 --version
```

2. **Add Python to path**

   To update the .bashrc file, you can choose between two methods.
   The first option is to open the .bashrc file directly in the
   VS Code text editor and paste the following line (shown in
   Code-block 2-22) at the end of the file.

   Code-block 2-22. Paste at the end of the .bashrc file

   ```
   alias python='usr/bin/python3.9'
   ```

Alternatively, you can make this change directly in the terminal using the command
shown in Code-block 2-23.

Code-block 2-23. Run it on the terminal

```
echo "alias python=''usr/bin/python3.9'" >> ~/.bashrc
```

After applying either method, ensure the changes take effect by running the
command in Code-block 2-24.

Code-block 2-24. Run it on the terminal

```
source ~/.bashrc
```

Both methods achieve the same result, so choose the one that works best for your
workflow. Ensure you execute the final step to make the alias immediately available.

Set Up and Activate a Virtual Environment

Navigate to the repository folder with cd ~, create a virtual environment with
python -m venv, and activate it with source; see Code-block 2-25.

Code-block 2-25. Move to the repository folder

```
cd ~
cd Documents/GitHub/deploy-secure-ds-apps-book
python -m venv venv-webs
source venv-webs/bin/activate
```

Deploy a Simple HTTP Streamlit App Through the EC2 Instance

Streamlit is a widely used open source framework designed specifically for building interactive dashboards and data visualizations. It simplifies the process of creating user-friendly interfaces for data science applications by allowing developers to write Python code directly. With Streamlit, there is no need to learn complex front-end web development frameworks or JavaScript; you can build fully interactive applications with just a few lines of Python.

This makes it a favorite among data scientists, analysts, and developers who want to quickly prototype and deploy visualizations and tools. Additionally, Streamlit's built-in features, such as real-time updates, sliders, and customizable widgets, make it a powerful choice for creating highly engaging applications that can integrate seamlessly with machine learning workflows.

To validate the setup, we will deploy a simple Streamlit app directly on the EC2 instance.

Steps to Deploy the App

1. **Install Streamlit**

   Ensure you are in the virtual environment and install Streamlit; refer to Code-block 2-26.

   Code-block 2-26. Connect to virtual env and pip install Streamlit

   ```
   cd ~/Documents/GitHub/deploy-secure-ds-apps-book
   source venv-webs/bin/activate
   pip install streamlit
   ```

2. **Create a Streamlit app**

   Create a new file named st_example.py in the repository folder
   and add the following code from Code-block 2-27. The code is
   also available at book's GitHub repository under the respective
   chapter.

   Code-block 2-27. Streamlit dummy example, st_example.py.

```python
import streamlit as st

def main():
    st.title("Streamlit App running ec2 via port 8501")
    st.write("Test successful!")

if __name__ == "__main__":
    main()
```

3. **Run the app**

   Then back to the terminal, run this Streamlit app; see
   Code-block 2-28 (make sure you're in the folder Documents/
   GitHub/deploy-secure-ds-apps-book).

   Code-block 2-28. Run streamlit st_example.py.

```
streamlit run st_example.py
```

4. **Access the application**

   In the terminal, you'll notice a couple of links provided to access
   the application. One will use localhost:8501, and the other will
   include the public IP address assigned through the Elastic IP setup
   in Chapter 1, section "Set Up an Elastic IP." For example, it might
   look like 54.94.167.229:8501. Here, 8501 is simply the port on
   which the application is running.

Summary

In this chapter, you have successfully prepared your EC2 instance environment for hosting secure and scalable data science applications. Specifically, you have

- Connected securely to your EC2 instance using SSH with VS Code, providing a robust and integrated development environment

    - VS Code Remote documentation: `https://code.visualstudio.com/docs/remote/remote-overview`

- Installed Docker and Docker Compose for standardized and reproducible container-based application deployments

    - Docker Installation documentation: `https://docs.docker.com/get-started/get-docker/`

- Configured GitHub access to manage your code repositories securely and efficiently, including SSH key generation and repository cloning

    - GitHub SSH Key Setup documentation: `https://docs.github.com/en/authentication/connecting-to-github-with-ssh`

- Set up Python and created a virtual environment to isolate and manage project-specific dependencies

    - Python Installation documentation: `https://docs.python.org/3/using/index.html`

- Deployed and validated a simple HTTP application using Streamlit directly on your EC2 instance

    - Streamlit documentation: `https://docs.streamlit.io/`

In the next chapter, you'll further enhance your application's security and scalability by setting up an AWS Load Balancer, creating target groups, adjusting security groups accordingly, and deploying a simple HTTP Streamlit application behind the load balancer.

## CHAPTER 3

Load Balancer on Your AWS Console

In the previous chapter, you established a robust development environment by securely accessing your EC2 instance through SSH with VS Code and installing essential tools such as Docker, Docker Compose, GitHub, and Python with a virtual environment. You also deployed a simple HTTP application using Streamlit. In this chapter, you'll further improve the security, scalability, and reliability of your infrastructure by introducing a load balancer. You'll learn to create and configure an AWS Network Load Balancer, establish multiple target groups for different application services, adjust security groups to manage secure access, and deploy a Streamlit application accessible through the load balancer.

Create a Load Balancer and the First Target Group

A load balancer serves as the gateway to a web server, exposed to the public internet to receive HTTP requests, and forwards them to target groups on the intranet that will then finally forward the request to the EC2 instance. So from that view, we can see the load balancer as an extra layer of security to your server, and also it helps to balance the amount of requests that might come at the same time so that your server is not overloaded.

That's an extra layer because we had already set up security groups in Chapter 1, section "Security Groups," that would allow only certain IP addresses to send HTTP requests to the server.

Also in some cases, if you're deploying an application within the company you work for, it might be that it's not even possible to deploy it using the instance's external IP as we did in Chapter 2, section "Deploy a Simple HTTP Streamlit App Through the EC2 Instance," and if that's the case, using a load balancer can be a great solution.

© Lucas H. Benevides e Braga 2025
L.H.B.e. Braga, *Deploying Secure Data Science Applications in the Cloud*,
https://doi.org/10.1007/979-8-8688-1715-1_3

See Figure 3-1 to have an overview of how the final infrastructure will look like at the end of Chapters 3 and 4.

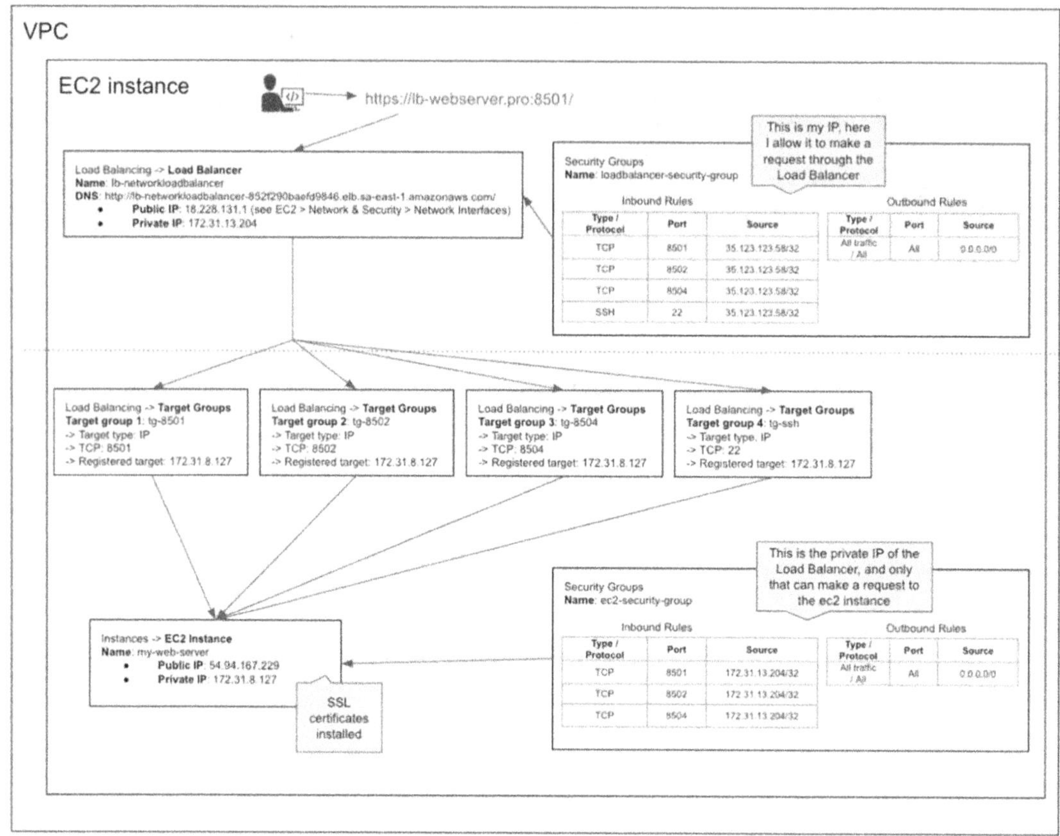

Figure 3-1. *Final infrastructure*

1. Navigate to **EC2 ➤ Load Balancers**, then click **Create load balancer**.

2. Select **Network Load Balancer** and click **Create**.

3. Under **Basic configuration**, provide the following details:

    - **Load balancer name:** Use a descriptive name such as lb-networkloadbalancer.

    - **Scheme:** Select **internet-facing** to expose the load balancer to public traffic.

    - **Load balancer IP address type:** Use the default option **IPv4**.

4. Under **Network mapping**:

- Select the same **VPC** used for the EC2 instance.

- **Availability zones**: Ensure it also matches the one from the EC2 instance. To verify, go to **EC2 ➤ Instances**, select your instance, and check the **Availability Zone** field.

Refer to Figure 3-2 for a visual guide to this setup.

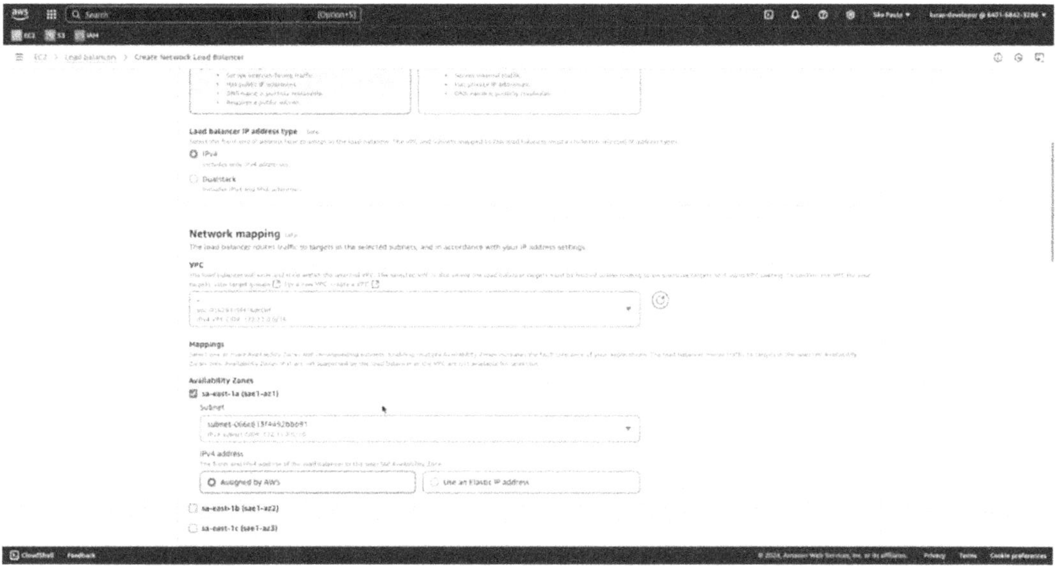

Figure 3-2. *Creating a Network Load Balancer*

5. Under **Security groups**, remove the default security group and add the one created for the EC2 instance: `instance-security-group`.

6. Under **Listeners and routing**, create the first target group:

- **Protocol**: TCP

- **Port**: 8501

- **Default action**: Create a target group (which will open another tab), step 1:

  - **Choose a target type**: IP addresses.

  - **Target group name**: Use a descriptive name such as `tg-8501`.

- **Protocol : Port**: It should be the same used for the listener, TCP and 8501.

- **IP address type**: Use the default option **IPv4**.

- **VPC**: Same VPC used for the EC2 instance.

- **Health checks**: Use the default TCP.

- Click **Next** to move to step 2.

  - **Network**: Use the same **VPC**.

  - **Enter an IPv4 address from a VPC subnet**: Enter the private IP of the EC2 instance. To verify, go to **EC2 ➤ Instances**, find your instance, and search for the field **Private IPv4 addresses**.

  - Click the button **Include as pending below** and click **Create target group**.

Refer to Figure 3-3 to visualize step 2 *Register targets* from creating a target group.

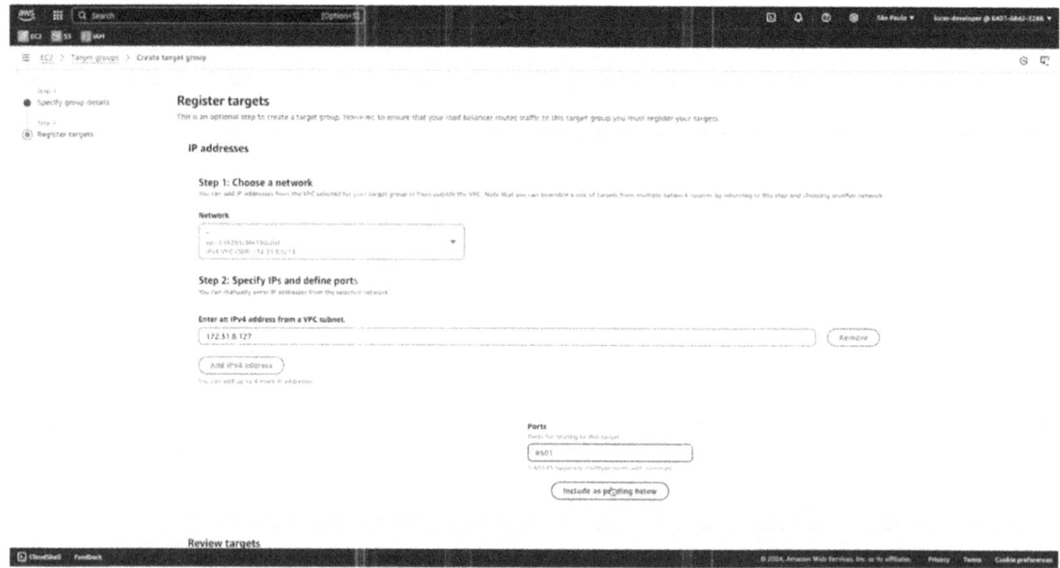

Figure 3-3. *Creating the first target group*

7. Go back to the previous tab **EC2 ➤ Load Balancers ➤ Create
   Network Load Balancer** and click refresh; it's a clockwise arrow.
   Then select the target group `tg-8501` that was created just now.

8. At the bottom right, click **Create load balancer.**

Create and Set Up Target Groups

The process of creating target groups for the load balancer is nearly the same as we did
in the section "Create a Load Balancer and the First Target Group." The difference is that
once the load balancer is already created, additional target groups can still be added
separately and then attached to the load balancer.

So far one target group was created, on port **8501** for the **Streamlit** application
(8501 is Streamlit's default port). This section covers the creation of the other following
target groups:

- **Port 22**: For **SSH connections** through the load balancer, enhancing
  security by avoiding direct instance access

- **Port 8502**: For a **Flask API**

- **Port 8504**: For the **Jenkins server**

1. Navigate to **EC2 ➤ Target groups**, then click **Create target group**
   (follow the same steps as from the section "Create a Load Balancer
   and the First Target Group").

   - **Choose a target type**: IP addresses.

   - **Target group name**: Use a descriptive name for each group (e.g.,
     `tg-ssh-22`, `tg-flask-8502`, `tg-jenkins-8504`).

   - **Protocol : Port**: Choose TCP, and enter the specific port for each
     service (e.g., 22, 8502, 8504).

   - Same steps as described in the previous section 'Create a Load
     Balancer and the First Target Group.'

- Click **Next** to move to step 2.

    - Same steps as described in the previous section 'Create a Load Balancer and the First Target Group'.

    - **Enter an IPv4 address from a VPC subnet**: Enter the private IP of the EC2 instance. To verify go to **EC2 ➤ Instances**, find your instance, and search for the field **Private IPv4 addresses**.

    - Click the button **Include as pending below** and click **Create target group**.

2. Go to **EC2 ➤ Load Balancers**; select your load balancer `lb-networkloadbalancer`.

3. Under the **Listener**, click **Add listener**.

4. For each port (e.g., 8502), configure the same way as was done in the section "Create a Load Balancer and the First Target Group":

   - **Protocol**: TCP

   - **Port**: For example, 8502

   - **Default action**: Select the corresponding target group (now already created) from the drop-down list (e.g., `tg-flask-8502`)

5. Then click the orange button (**Add**) at the bottom right to finalize the action.

Refer to Figure 3-4 to visualize the page where you find the **Add listener** option (note the path on the top left **EC2 ➤ Load Balancers ➤ lb-networkloadbalancer**).

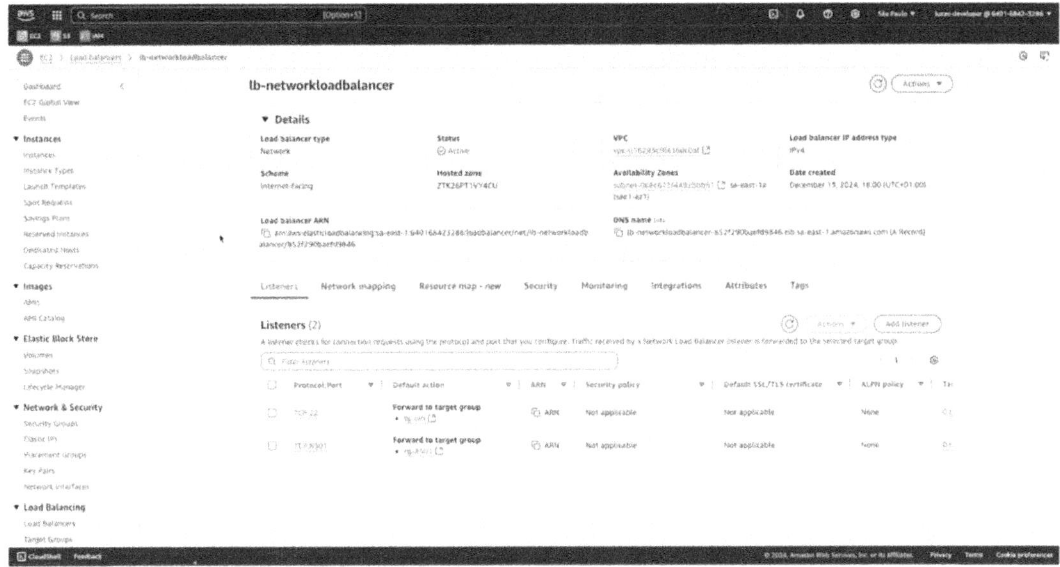

Figure 3-4. *Add listener*

After all this process of adding target groups, the final state should look like what's displayed in Figure 3-5.

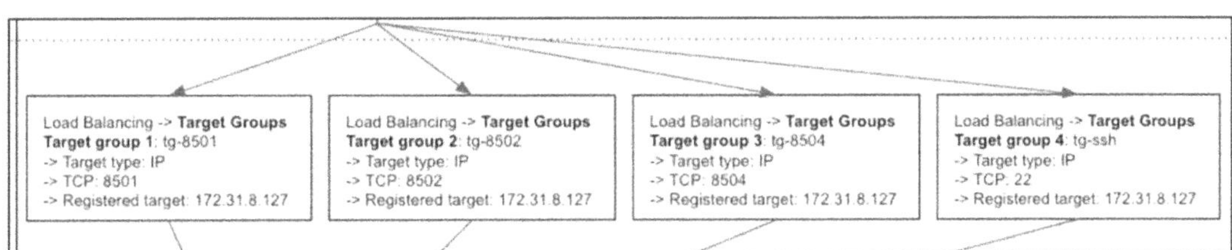

Figure 3-5. *Target groups final state*

Adjust Security Groups

To secure your infrastructure, the security groups must be reorganized. Firstly so that the **load balancer** faces the public internet but only allows specific IP addresses to access it. You can choose for instance an IP that belongs to a VPN network, and only people that connect to that VPN will be able to make a request to the load balancer, hence to the web server. The security group for the load balancer was named `loadbalancer-security-group`. Secondly, the EC2 instance is accessible only through the load balancer, with no direct public exposure. The security group for the EC2 instance was named `ec2-security-group`.

Refer to the section "Configure Security Groups" in Chapter 1 to remember how to set up the security groups; in the end, the `loadbalancer-security-group` should look like Figure 3-6, and the `ec2-security-group` should look like Figure 3-7.

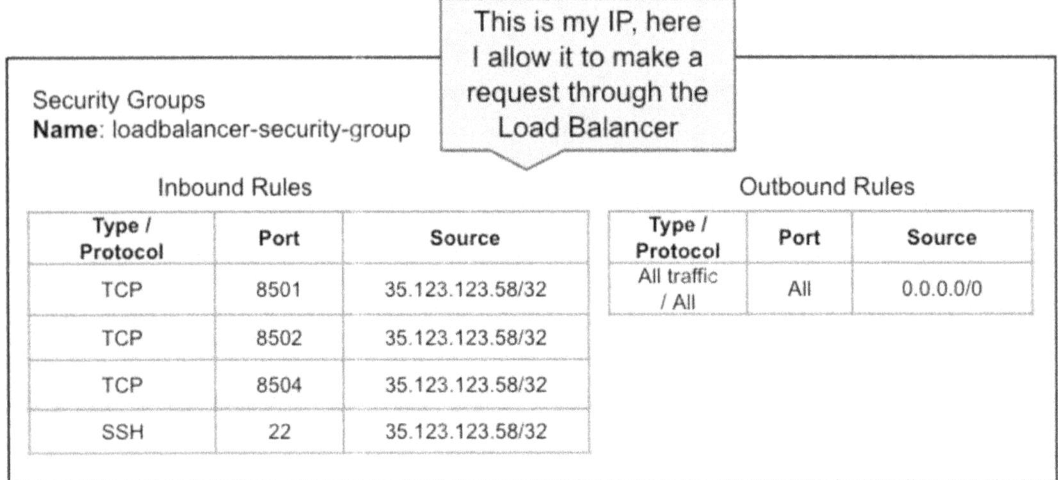

Figure 3-6. *Load balancer's security group (subset of Figure 3-1)*

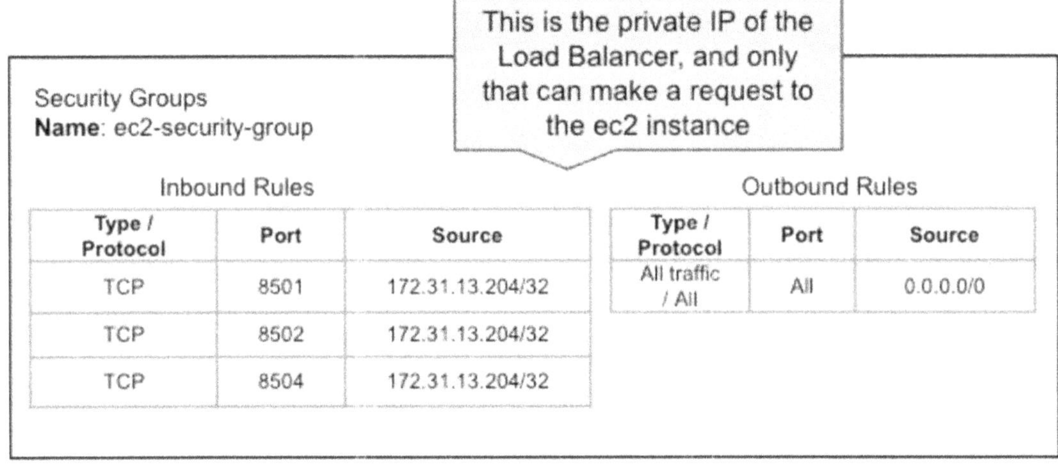

Figure 3-7. *EC2 instance's security group (subset of Figure 3-1)*

Deploy a Simple HTTP Streamlit App Through the Load Balancer

With the security groups and target groups in place, we can now deploy a simple Streamlit application accessible through the load balancer.

Steps to Deploy an HTTP Streamlit App Through the Load Balancer

1. Similar to what was done in the section "Deploy a Simple HTTP Streamlit App Through the EC2 Instance" in Chapter 2. **Run the Streamlit app on the EC2's terminal, as shown in Code-block 3-1.**

   Code-block 3-1. Deploy Streamlit on the terminal

   ```
   cd ~/Documents/GitHub/deploy-secure-ds-apps-book
   source venv-webs/bin/activate
   streamlit run st_example.py
   ```

2. **Access** the app **through the load balancer**'s DNS that was provided by AWS, but adding the port 8501 at the end.

   ```
   lb-networkloadbalancer-852f290baefd9846.elb.sa-east-1.amazonaws.
   com:8501
   ```

Figure 3-8 shows how the Streamlit app looks like once you access it.

Figure 3-8. Accessing Streamlit through the load balancer

Summary

You have significantly improved your application's security and scalability by implementing a load balancer with the appropriate security settings. You have

- Created a Network Load Balancer to be the front door of your application and distribute incoming traffic effectively and securely

    - AWS documentation for Network Load Balancer and target groups: `https://docs.aws.amazon.com/elasticloadbalancing/latest/network/introduction.html`

- Established target groups to route the incoming requests to the corresponding applications

- Adjusted security groups to ensure

  - The load balancer securely interfaces with the public internet, restricted to specific IP addresses or VPNs

  - The EC2 instance is isolated from direct public exposure, receiving traffic exclusively via the load balancer

- AWS documentation for security groups: `https://docs.aws.amazon.com/vpc/latest/userguide/vpc-security-groups.html`

- Deployed and accessed a simple **HTTP Streamlit application** via the load balancer, validating your configuration

In the next chapter, you'll continue to improve your infrastructure by acquiring and configuring a domain name and SSL certificates to enable secure HTTPS connections. You will learn how to purchase a domain, set up an SSL certificate, configure your infrastructure accordingly, and deploy an HTTPS-secured application accessible via your custom domain name.

## CHAPTER 4

Domain Name and SSL Certificates

In the previous chapter, you significantly improved your infrastructure by configuring an AWS Network Load Balancer, setting up multiple target groups, adjusting security groups, and deploying a simple HTTP Streamlit application accessible through the load balancer. In this chapter, you'll build upon that infrastructure by integrating a professional domain name and SSL certificates, ensuring secure encrypted HTTPS traffic. You will learn how to purchase and configure a domain and SSL certificate, route traffic securely through your domain, and deploy an HTTPS-secured Streamlit application using Docker and Nginx.

Buy the Domain Name and SSL Certificates Through namecheap.com

The first step is to acquire a domain name and SSL certificates, and namecheap.com is a user-friendly and reliable platform for purchasing both.

Steps to Purchase

1. Visit namecheap.com and search for a domain name that suits your application; see Figure 4-1.

   - So if your domain preference is lb-webserver.com, try only lb-webserver and see all the options available. Usually .com is more expensive, and there are options like .pro or .xyz that sometimes cost only about 3$ for the first year.

© Lucas H. Benevides e Braga 2025
L.H.B.e. Braga, *Deploying Secure Data Science Applications in the Cloud*,
https://doi.org/10.1007/979-8-8688-1715-1_4

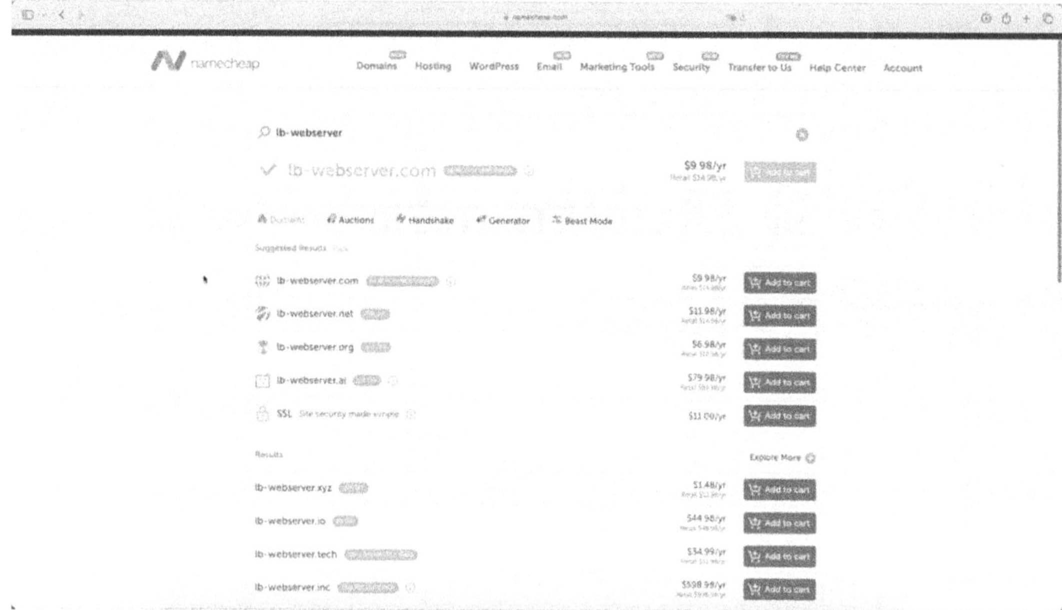

Figure 4-1.   *Visit namecheap.com*

2. Add the chosen domain to your cart and include an SSL certificate in your purchase.

    •   There are different SSL certificate options; however, for now, the basic option Positive SSL is enough; see Figure 4-2.

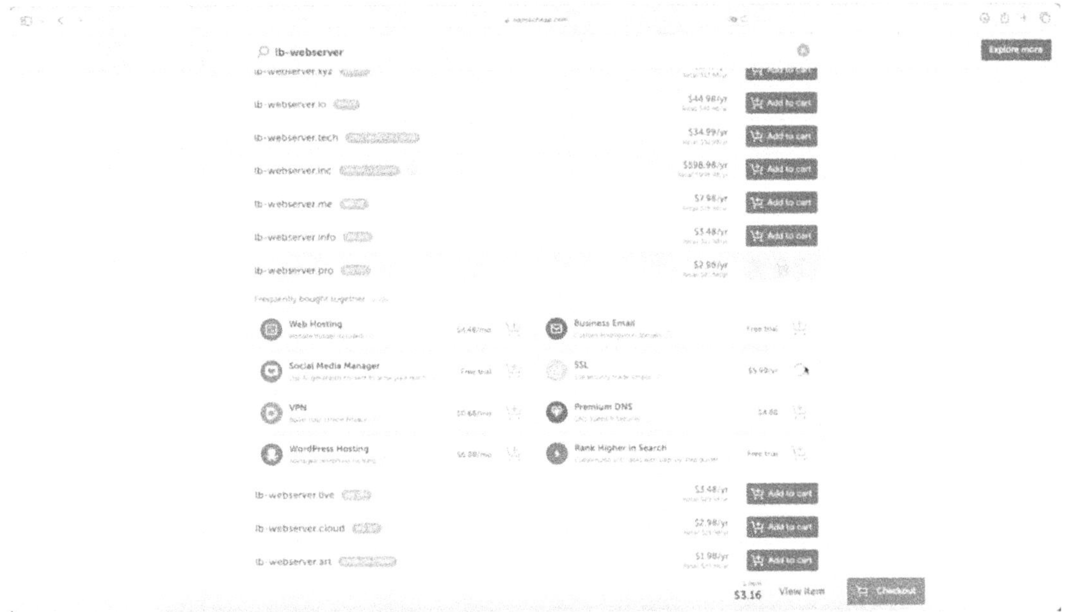

Figure 4-2. *Add SSL certificate to the cart*

3. Proceed to checkout; see Figure 4-3.

- Double-check that you have a Domain registration for 1 year (or more in case you'd want it; however, you can opt to renew it later as well). And double-check that you have a PositiveSSL certificate on the basket as well.

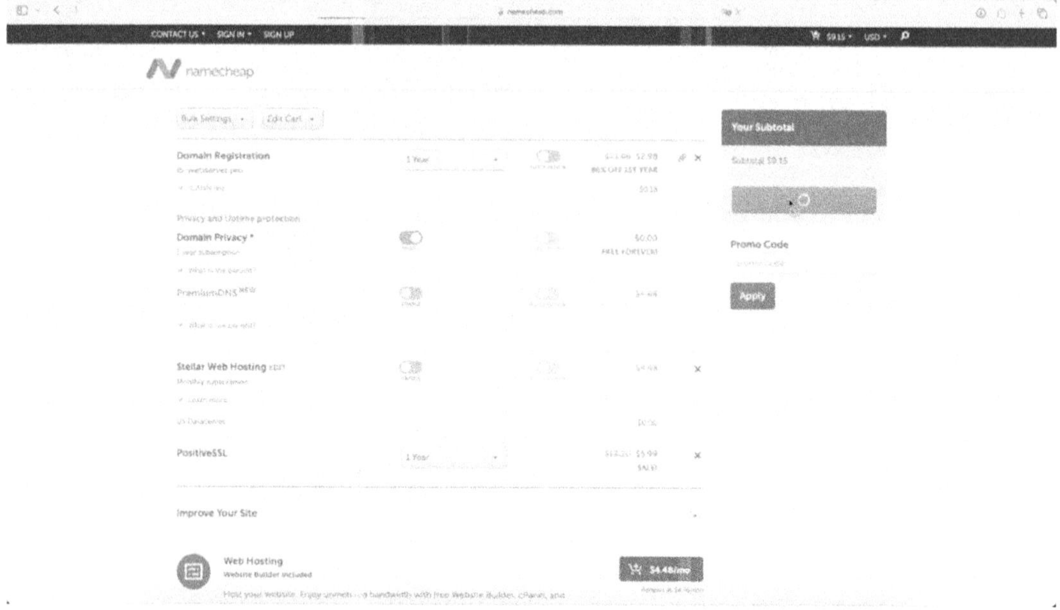

Figure 4-3. *Proceed to checkout*

Once purchased, the domain and SSL certificates will be accessible from your Namecheap account dashboard.

Set Up the SSL Certificates

SSL certificates ensure secure communication between your server and its clients. In this section, you will generate a certificate signing request (CSR), activate your SSL certificate, and configure your domain's DNS settings.

Generate a Certificate Signing Request (CSR)

1. Go to VS Code, and connect to your EC2 instance via SSH.

    On the terminal of the EC2 instance (on VS Code), navigate to the home folder and run the following commands (adjust file names as needed, e.g., `lb-webserver_pro_csr.crt` instead of your_domain.crt) from Code-block 4-1, also shown in Figure 4-4.

Code-block 4-1. Create certificate request

```
cd ~


openssl req -new -newkey rsa:2048 -nodes -keyout private.key -out
your_domain_csr.crt
```

Figure 4-4. *Generate a certificate signing request (CSR)*

2. Fill in the prompted fields (press enter after each field), such as

   a. **Country name:** Enter your country code (e.g., DE for Germany, United States is US, Brazil is BR, etc.).

   b. **State/province:** Your state or province (e.g., Berlin, New York).

   c. **Locality name:** Your city within that state/province (e.g., Berlin, New York, Fortaleza)

   d. **Organization name:** Use your company's/organization's name or a domain name.

   e. **Common name:** Enter the domain name you purchased (e.g., example.com, lb-webserver.pro).

   f. **Email address:** Enter your company's/organization's email address or your personal email address.

   g. **A challenge password:** You can leave that blank and press enter or you can create a password and press enter.

   h. **An optional company name:** Similar to organization name, it's optional and of your choosing.

3. Save the generated `.crt` file for the next step; see Figure 4-5. Notice that within that same folder that you ran those commands, your home folder (`cd ~`), these two files were created:

   a. `private.key`, save it on this folder `~/Documents/GitHub/my-web-server/config/private/`

   b. `your_domain_csr.crt`, save it on this folder `~/Documents/GitHub/deploy-secure-ds-apps-book/config/`

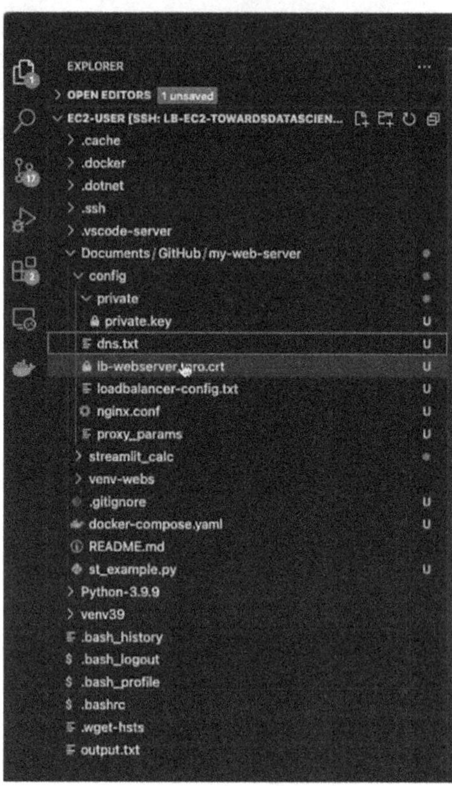

Figure 4-5. *Save CSR and private key*

Activate the SSL Certificate: From ACTIVATE to PENDING

1. Open the file `your_domain_csr.crt` that was just created, and then **copy** all of its content; see Figure 4-6.

2. Log in to Namecheap and navigate to **SSL Certificates** (left panel).

3. Click **Activate** next to the certificate you purchased.

4. **Paste** the contents of the your_domain_csr.crt file into the **CSR field**; see Figure 4-7.

5. Check that on the Primary domain field your domain name will be displayed (e.g., lb-webserver.pro), and click **Next**.

   - The Primary domain field will be filled out automatically with the domain that you used to generate the certificate signing request (CSR).

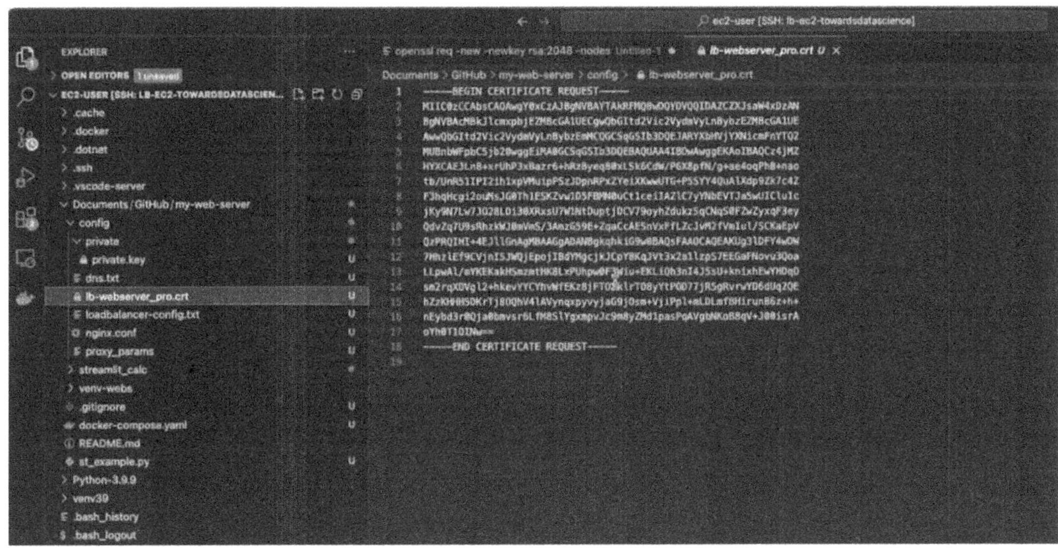

Figure 4-6. *Copy the content of the CSR file*

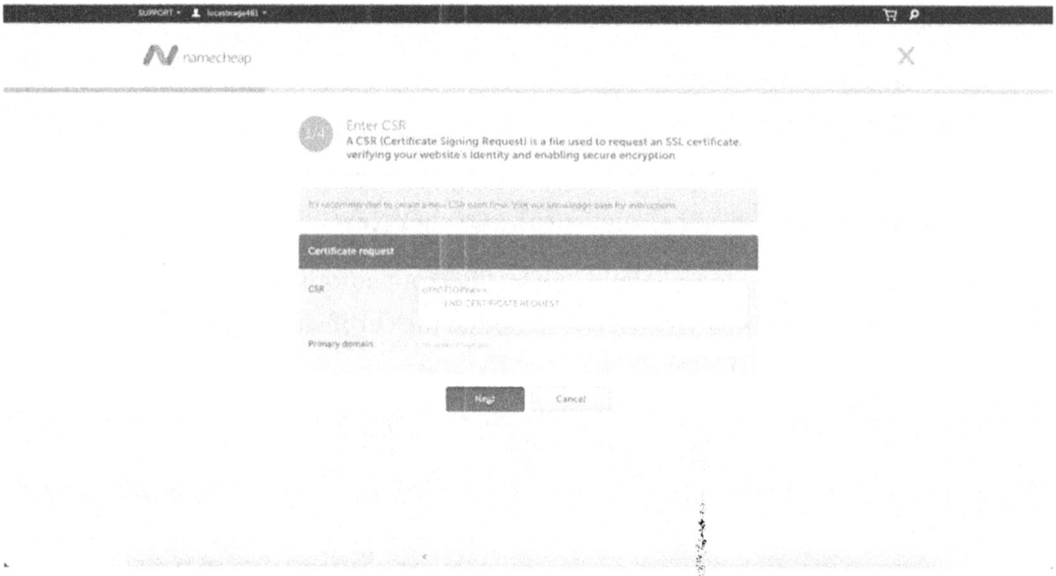

Figure 4-7. *Paste the content of the CSR file into Namecheap*

6. Select the validation **method**: add CNAME record and press **Next**.

7. **Confirm** the **recipient email** to receive the SSL files; it will be filled out automatically with the email that you used to generate the certificate signing request (CSR). Press **Next** and then press **Submit**.

Activate the SSL Certificate: From PENDING to ACTIVE

1. Still on namecheap.com, on the left panel, click **SSL Certificates**; you will see that ACTIVATE is no longer available and that the status has changed to PENDING; see Figure 4-8.

2. Click **DETAILS**.

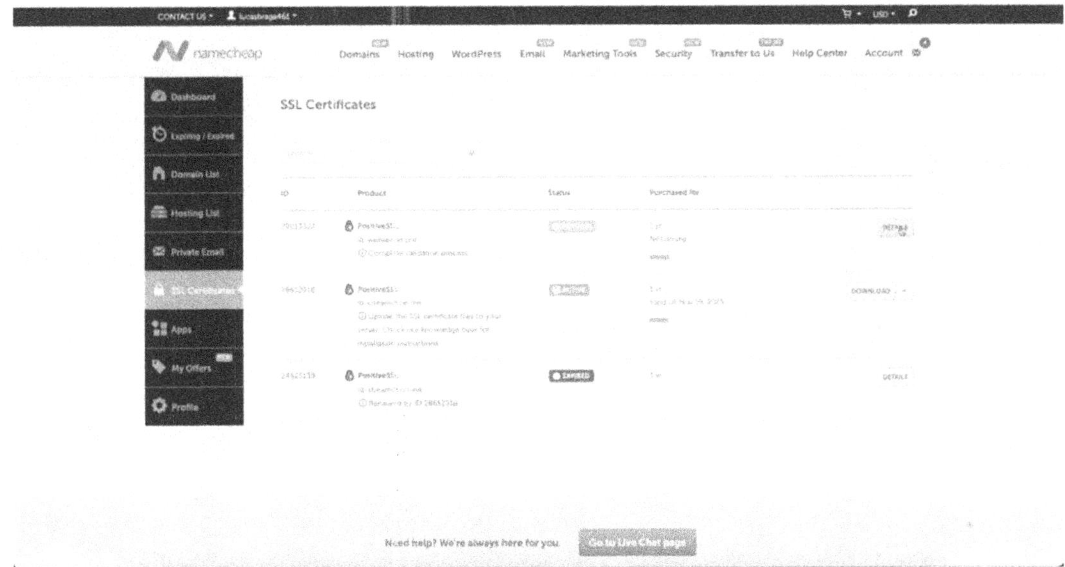

Figure 4-8. *Status is now PENDING*

3. After clicking DETAILS, scroll all the way down and click **See CSR code** (under Admin details).

4. The Certificate activation has initiated; the Certificate Status is IN PROGRESS; now on **EDIT METHODS**, click the drop-down, and click **Get Record**. A popup opens showing a host and a target key; this popup is referred to here as *Get Record tab*.

5. Copy the host key without the suffix (i.e., without the domain); see Figure 4-9.

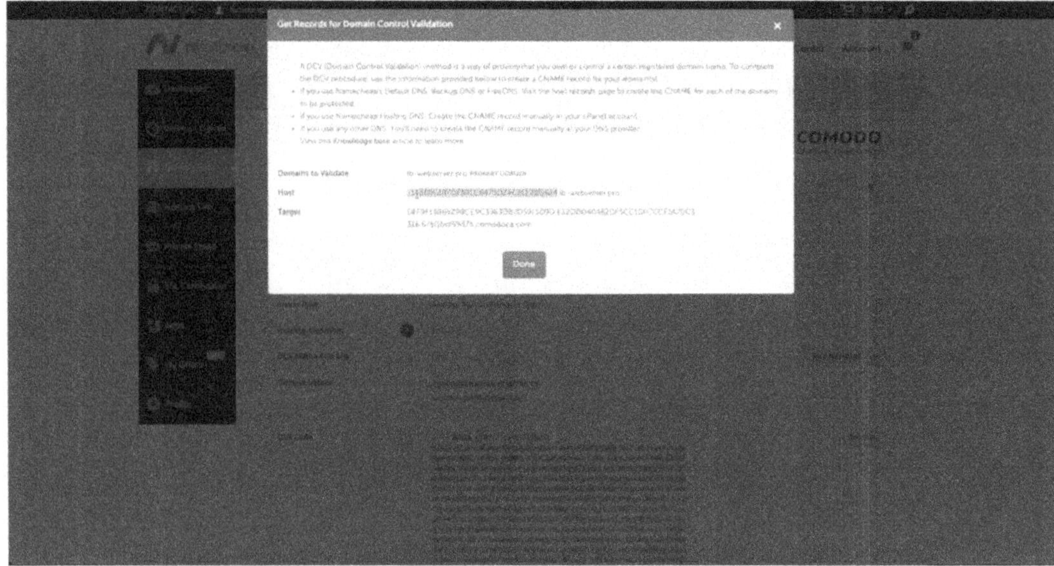

Figure 4-9. *On the Get Record tab, copy the host key*

6. On a different tab, still on namecheap.com, go to **Domain List**,
    find your domain, click **MANAGE**, and then click **Advanced DNS**;
    this tab is referred to here as *Advanced DNS tab* (see Figure 4-10).

7. If there is already a field **CNAME Record**, then edit it (otherwise
    add it). Replace its host by the host key that you copied from step 5
    (otherwise add a CNAME record); see Figure 4-10.

8. Back to the *Get Record tab*, where you copied the host key, copy
    also the target (the entire text, including suffix; see Figure 4-11).

9. And again back to the *Advanced DNS tab*, paste the **target key** on
    the **field Value**.

10. Change the *TTL* field to the minimum value (1 min), and click the
    green check mark ✓ to save changes.

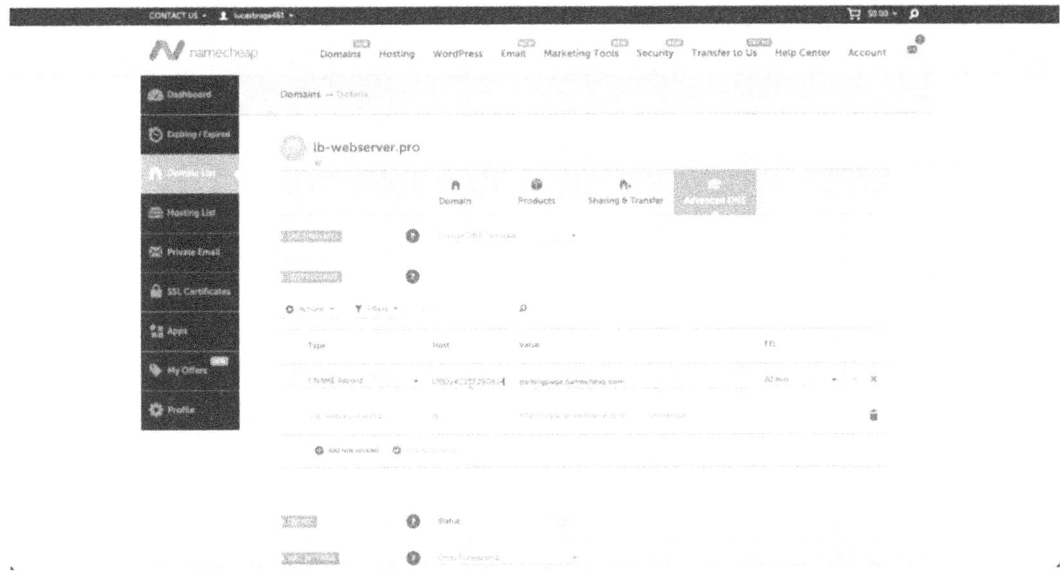

Figure 4-10. *Advanced DNS tab; paste the host key on the host field*

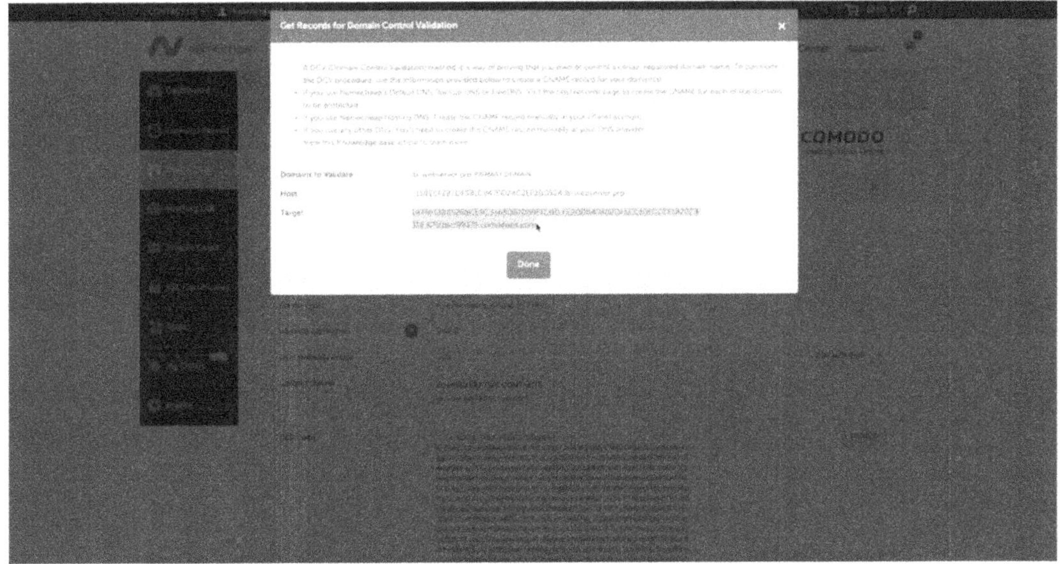

Figure 4-11. *Copy the target key*

Go back to the **SSL Certificates** page (left panel); refresh the page; the status should be now set as **ACTIVE**. And you will receive an email with the certificate files.

Route traffic to your domain.

These steps are to set up the access of your web server through the domain instead of the IP address.

1. Back to the *Advanced DNS tab* (tab you see from Figure 4-10), click **ADD NEW RECORD**:

   - **Type: A Record**

   - **Host: @**

   - **Value: <load_balancers_public_ip>**. You can find your load balancer's Public IP on the AWS console; navigate to **EC2 ➤ Network & Security ➤ Network Interfaces** and you will find one record for the load balancer and one record for the EC2 instance. Identify which one is your load balancer and look up the field **Public IPv4 address** (e.g., 18.228.131.1).

   - **TTL: 1 min** (or the lowest value you can find), and click the green check mark ✓ to save changes.

See Figure 4-12.

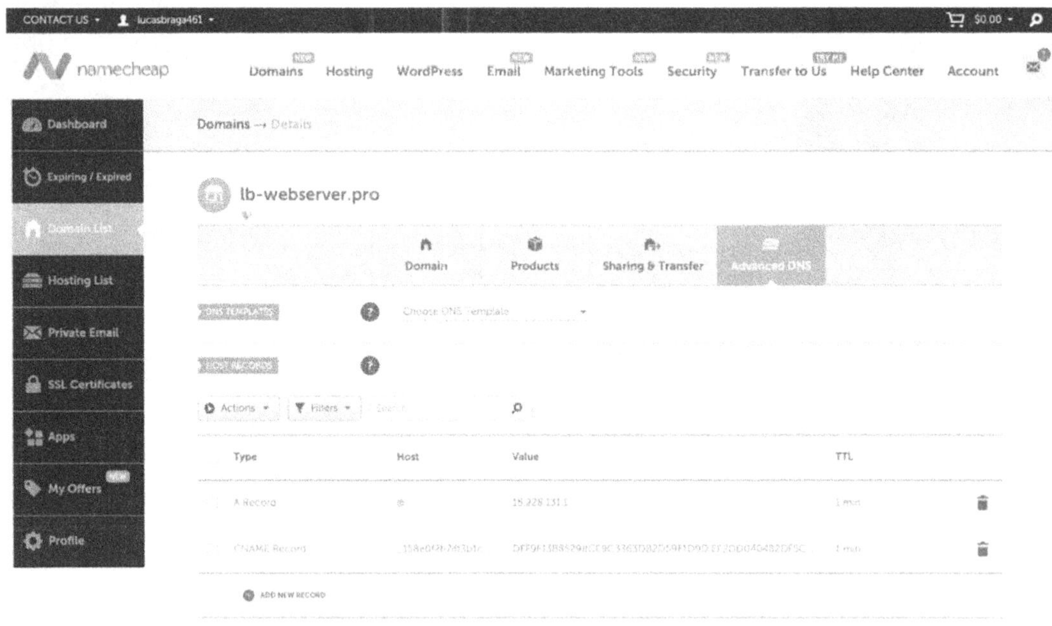

Figure 4-12. *Load balancer mapped to the domain*

Prepare the Certificates

By now, you should have received an email containing the certificate files. If not, you can also download them directly from the **SSL Certificates** section in the left panel of the Namecheap console.

1. Once you have the `.zip` file with the certificates, begin by unzipping it.

2. After extracting the files, upload them to the `config` folder within your VS Code repository. (In this guide, the unzipped folder is named `lb-webserver_pro`.)

3. Combine the certificates into a single file:

    •   Create a file within the config folder and name it like `your_domain_combined.crt` (e.g., `lb-webserver_pro_combined.crt`).

    •   Go to the recently unzipped folder, copy the contents of the file `your_domain.crt` into `your_domain_combined.crt` (be careful not to confuse `your_domain.crt` with the certificate signing request, CSR; they are different files).

    •   Go back to the recently unzipped folder, copy the contents of the file `your_domain.ca-bundle`, and back to `your_domain_combined.crt`, add a new line at the bottom and paste the content.

    •   At the end, organize the files within the config folder so that it contains only the `your_domain_combined.crt` and the `private_key` which will go within a folder named private within config, like `config/private/private.key`; see Figure 4-13.

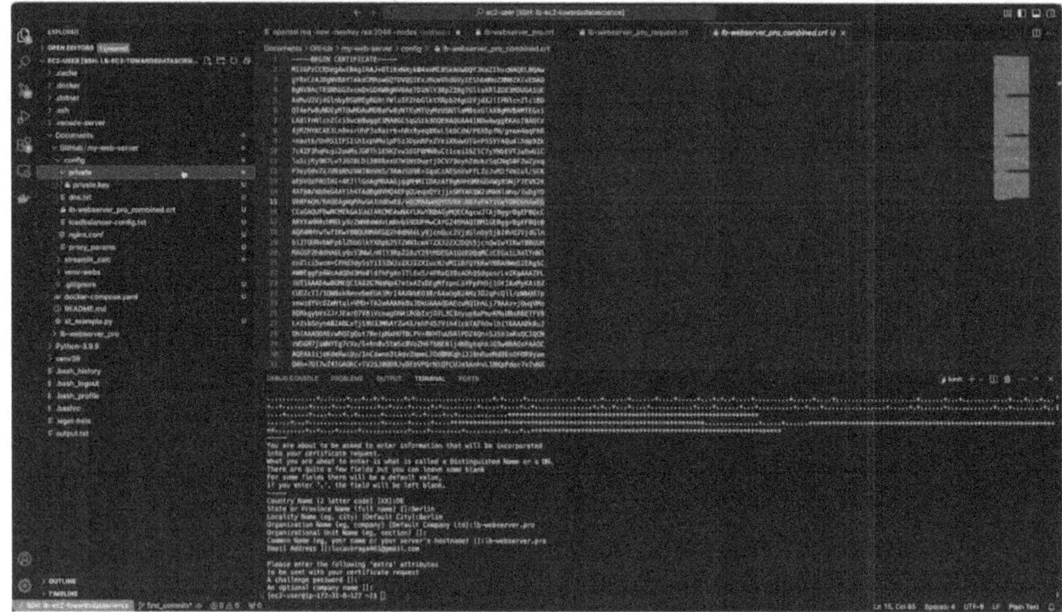

Figure 4-13. *Config folder structured*

Deploy an HTTPS Streamlit App with Domain Name

Now, let's perform another deployment to test whether the domain and SSL certificates are working. While this isn't the final deployment, it will serve as a validation step. For this test, we'll deploy only the Streamlit dashboard using a Docker container. To do so, we need to modify two configuration files by either commenting out or temporarily removing the lines related to other applications.

1. **Modify the `docker-compose.yaml` file**

   Update the `docker-compose.yaml` file so it matches the configuration shown in Code-block 4-2. This updated file is also available in the GitHub repository: docker-compose-test.yaml.

   Code-block 4-2. Example `docker-compose.yaml` file

   ```
   version: "3.7"

   services:
    streamlit_calc:
   ```

```
    build:
      context: streamlit_calc
    ports:
      - "8501"
    restart: always

 nginx:
    image: nginx:latest
    volumes:
      - ./config/nginx.conf:/etc/nginx/nginx.conf:ro
      - ./config/lb-webserver_pro_combined.crt:/etc/pki/nginx/lb-
        webserver_pro_combined.crt:ro
      - ./config/private/private.key:/etc/pki/nginx/private/
        private.key:ro
      - ./config/proxy_params:/etc/nginx/proxy_params:ro
    depends_on:
      - streamlit_calc
    ports:
      - "8501:8501"
    restart: always
```

2. **Modify the `config/nginx.conf` file**

   Update the `nginx.conf` file located in the `config` folder. Adjust it
   to match the configuration shown in Code-block 4-3. This version
   is also available in the GitHub repository: nginx-test.conf.

Code-block 4-3. Example `nginx.conf` file

```
user nginx;
worker_processes auto;
error_log /var/log/nginx/error.log;
pid /run/nginx.pid;

# Load dynamic modules. See /usr/share/doc/nginx/README.dynamic.
include /usr/share/nginx/modules/*.conf;
events {
    worker_connections 1024;
}
```

```
http {
    log_format  main  '$remote_addr - $remote_user [$time_local]
    "$request" '
                        '$status $body_bytes_sent "$http_referer" '
                        '"$http_user_agent" "$http_x_forwarded_for"';

    access_log  /var/log/nginx/access.log  main;

    sendfile            on;
    tcp_nopush          on;
    tcp_nodelay         on;
    keepalive_timeout   65;
    types_hash_max_size 4096;
    include             /etc/nginx/mime.types;
    default_type        application/octet-stream;
    include /etc/nginx/conf.d/*.conf;

    server {
        listen       80;
        listen       [::]:80;
        server_name  _;
        root         /usr/share/nginx/html;

        include /etc/nginx/default.d/*.conf;
        error_page 404 /404.html;
        location = /404.html {
        }
        error_page 500 502 503 504 /50x.html;
        location = /50x.html {
        }
    }
    server {
        listen 80;
        listen [::]:80;
        server_name lb-webserver.pro www.lb-webserver.pro;
        return 301 https://$server_name$request_uri;
    }
```

```
# Settings for a TLS enabled server.
   server {
       listen 8501 ssl;
       server_name lb-webserver.pro www.lb-webserver.pro;
       root           /usr/share/nginx/html;

       ssl_certificate "/etc/pki/nginx/lb-webserver_pro_
       combined.crt";
       ssl_certificate_key "/etc/pki/nginx/private/private.key";
       ssl_session_cache shared:SSL:1m;
       ssl_session_timeout  10m;
       ssl_prefer_server_ciphers on;

       include /etc/nginx/default.d/*.conf;
       error_page 404 /404.html;
           location = /40x.html {
       }

       error_page 500 502 503 504 /50x.html;
           location = /50x.html {
       }

       location / {
           include proxy_params;
           proxy_pass http://streamlit_calc:8501;
           proxy_http_version 1.1;
           proxy_set_header Upgrade $http_upgrade;
           proxy_set_header Connection "upgrade";
           proxy_set_header Host $host;
           proxy_set_header X-Real-IP $remote_addr;
           proxy_set_header X-Forwarded-For $proxy_add_x_
           forwarded_for;
           proxy_set_header X-Forwarded-Proto $scheme;
       }
   }
}
```

Once these changes are made, open VS Code and navigate to the **Extensions** section. Install the **Docker extension** by Microsoft. After installation, you'll see the Docker extension icon in the left panel. Navigate to the repository's main folder, then run the docker-compose command to deploy the application. Refer to Code-block 4-4 for the exact terminal commands.

Code-block 4-4. Run docker-compose

```
cd ~/Documents/GitHub/deploy-secure-ds-apps-book
docker-compose up -d –build
```

Verify the deployment by visiting your domain (e.g., `https://<your_domain>.com:8501`, `https://lb-webserver.pro:8501`) in a browser.

Summary

You have now established a secure, professional-grade web application environment. Specifically, you have

- Purchased a domain name and SSL certificates using Namecheap

- Generated a certificate signing request (CSR) and activated your SSL certificates, securing your application with encryption

    - Documentation on Namecheap SSL activation: `https://www.namecheap.com/support/knowledgebase/article.aspx/794/67/how-do-i-activate-an-ssl-certificate/`

- Configured DNS settings to route your domain traffic securely to your AWS Load Balancer

- Deployed a secure HTTPS Streamlit application using Docker and Nginx, accessible through your custom domain

    - Nginx SSL Configuration documentation: `https://docs.nginx.com/nginx/admin-guide/security-controls/terminating-ssl-http/`

    - Docker Compose documentation: `https://docs.docker.com/compose/`

In the next chapter, you'll continue expanding your infrastructure's capabilities by deploying more robust applications like Jenkins and Flask. You will enable Docker services to automatically start upon system boot. You'll also train a machine learning model and deploy this trained model for real-time predictions with Flask on your infrastructure.

## CHAPTER 5

Deploying More Robust Applications (Jenkins, Flask, and Streamlit)

In this chapter, we will deploy more robust and functional applications, including Jenkins for automation, Flask for API calls, and Streamlit for dashboards. These deployments will build on the foundation laid in earlier chapters by utilizing Docker, Nginx, and SSL certificates. The chapter also covers critical steps like enabling Docker services to start automatically, preparing VS Code for Jupyter notebooks, and creating a Flask application with authentication.

All code used in this chapter is available in the GitHub repository: `https://github.com/lucasbraga461/deploy-secure-ds-apps-book`. Clone the repository to access the files, which require minimal adjustments such as adding SSL certificates (as seen in Chapter 4) and modifying configuration files such as config/nginx.conf and docker-compose.yaml to match your setup.

Deploy Jenkins

Jenkins is a widely used automation server that facilitates continuous integration and delivery (CI/CD) pipelines. We will deploy Jenkins on port 8504 and perform its initial configuration.

In the `docker-compose.yaml` file, you'll notice a section for volumes within the Jenkins configuration. This volume maps the instance's local folder `jenkins/var/jenkins_home` to the container's internal folder `/app/var/jenkins_home`. This setup is crucial because when you run Docker Compose, Jenkins will generate files within the

© Lucas H. Benevides e Braga 2025
L.H.B.e. Braga, *Deploying Secure Data Science Applications in the Cloud*,
https://doi.org/10.1007/979-8-8688-1715-1_5

container. By syncing these files to the instance's local storage, you ensure that Jenkins retains its data even if you need to stop and restart the Docker container (a common scenario in development environments).

Steps to Deploy Jenkins

1. **Clone the repository**

   - Ensure you have cloned the repository into your EC2 instance (see the section "Set Up GitHub and Clone the Repository" in Chapter 2 for reference on cloning a GitHub repository).

2. **Adjust permissions**

   - Docker and other users like Nginx require proper permissions to access and execute commands within the Jenkins directory. Use the commands in Code-block 5-1 to set these permissions.

   Code-block 5-1. Linux commands to create users and grant permissions

```
# 1. create shared group
cd ~
sudo groupadd jenkins_shared

# 2. add ec2-user to the group
sudo usermod -aG jenkins_shared ec2-user

# 3. if docker already exist then:
sudo usermod -aG jenkins_shared docker

# otherwise:
sudo useradd -g jenkins_shared docker
sudo usermod -aG jenkins_shared docker

# 4. create jenkins user and add to jenkins_shared
sudo groupadd jenkins
sudo useradd -g jenkins jenkins
sudo usermod -aG jenkins_shared jenkins

# 5. create nginx user and add to jenkins_shared
sudo groupadd nginx
```

```
sudo useradd -g nginx nginx
sudo usermod -aG jenkins_shared nginx

# 6. change ownership of the jenkins directory
cd ~/Documents/GitHub/deploy-secure-ds-apps-book/jenkins/
sudo chown -R $(id -u jenkins):jenkins_shared var/jenkins_home/

sudo chmod -R 775 var/jenkins_home/

# 7. verify that they all belong to jenkins_shared
groups jenkins
groups docker
groups nginx
# If not add them
sudo usermod -aG jenkins_shared jenkins
sudo usermod -aG jenkins_shared docker
sudo usermod -aG jenkins_shared nginx

# 8. Create a .env file with the UID:GID from jenkins
#  on the same folder as the docker-compose.yaml file
#  these variables will be used by the docker-compose.yaml file
cd ~/Documents/GitHub/deploy-secure-ds-apps-book/
echo "JENKINS_UID=$(id -u jenkins)" > .env
echo "JENKINS_GID=$(id -g jenkins)" >> .env
```

3. **Run docker-compose**

- At this point, you should be ready to navigate to the repository's main folder and run another docker-compose command. If you prefer to proceed incrementally, adjust the docker-compose.yaml and nginx.conf files to include only Jenkins and Streamlit, temporarily excluding Flask. You can comment out the lines corresponding to Flask (see Code-block 5-2 and Code-block 5-3), as its deployment will be covered in sections "Enable Docker Service to Start Automatically" and "Deploy a Flask App for API Calls with Authentication."

Code-block 5-2a. Run docker-compose

```
cd ~/Documents/GitHub/deploy-secure-ds-apps-book
docker-compose up -d --build
```

Code-block 5-2b. Commenting flask out of docker-compose.yaml

```yaml
version: "3.7"

services:
 streamlit_calc:
   build:
     context: streamlit_calc
   ports:
     - "8501"
   restart: always

 # flask:
 #   build:
 #     context: flask
 #   ports:
 #     - "8502"
 #   volumes:
 #     - ./flask/logs:/app/logs
 #   restart: always

 jenkins:
   build:
     context: jenkins
   ports:
     - "8080"
   user: "${JENKINS_UID}:${JENKINS_GID}"
   environment:
     - JENKINS_UID=${JENKINS_UID}
     - JENKINS_GID=${JENKINS_GID}
     - AWS_CONFIG_FILE=/app/.aws/config
```

```
    volumes:
      - ./streamlit_calc/data:/app/data
      - ./jenkins/.aws:/app/.aws
      - ./jenkins/var/jenkins_home:/var/jenkins_home
    restart: always

  nginx:
    image: nginx:latest
    volumes:
      - ./config/nginx.conf:/etc/nginx/nginx.conf:ro
      - ./config/lb-webserver_pro_combined.crt:/etc/pki/nginx/
        lb-webserver_pro_combined.crt:ro
      - ./config/private/private.key:/etc/pki/nginx/private/
        private.key:ro
      - ./config/proxy_params:/etc/nginx/proxy_params:ro
    depends_on:
      - streamlit_calc
      - jenkins
      # - flask
    ports:
      - "8501:8501"
      - "8504:8504"
      # - "8502:8502"
    restart: always
```

Code-block 5-3. Commenting flask out of nginx.conf

```
user nginx;
worker_processes auto;
error_log /var/log/nginx/error.log;
pid /run/nginx.pid;

# Load dynamic modules. See /usr/share/doc/nginx/README.
dynamic.
include /usr/share/nginx/modules/*.conf;
```

```
events {
    worker_connections 1024;
}

http {
    log_format  main  '$remote_addr - $remote_user [$time_local]
    "$request" ''$status $body_bytes_sent "$http_referer" '
    '"$http_user_agent" "$http_x_forwarded_for"';

    access_log  /var/log/nginx/access.log  main;
    # proxy_headers_hash_max_size 1024;
    # proxy_headers_hash_bucket_size 128;

    sendfile            on;
    tcp_nopush          on;
    tcp_nodelay         on;
    keepalive_timeout   65;
    types_hash_max_size 4096;

    include             /etc/nginx/mime.types;
    default_type        application/octet-stream;

    # Load modular configuration files from the /etc/nginx/
    conf.d directory.
    # See http://nginx.org/en/docs/ngx_core_module.html#include
    # for more information.
    include /etc/nginx/conf.d/*.conf;

    server {
        listen      80;
        listen      [::]:80;
        server_name _;
        root        /usr/share/nginx/html;

        # Load configuration files for the default server block.
        include /etc/nginx/default.d/*.conf;
```

```
        error_page 404 /404.html;
        location = /404.html {
        }

        error_page 500 502 503 504 /50x.html;
        location = /50x.html {
        }
    }
    server {
        listen 80;
        listen [::]:80;
        server_name lb-webserver.pro www.lb-webserver.pro;
        return 301 https://$server_name$request_uri;
    }

# Settings for a TLS enabled server.
    server {
        listen 8501 ssl;
        server_name lb-webserver.pro www.lb-webserver.pro;
        root         /usr/share/nginx/html;

        ssl_certificate "/etc/pki/nginx/lb-webserver_pro_
        combined.crt";
        ssl_certificate_key "/etc/pki/nginx/private/
        private.key";
        ssl_session_cache shared:SSL:1m;
        ssl_session_timeout  10m;
        # ssl_ciphers PROFILE=SYSTEM;
        ssl_prefer_server_ciphers on;

        # Load configuration files for the default server block.
        include /etc/nginx/default.d/*.conf;

        error_page 404 /404.html;
            location = /40x.html {
        }
```

```
        error_page 500 502 503 504 /50x.html;
            location = /50x.html {
        }

    location / {
            include proxy_params;
            proxy_pass http://streamlit_calc:8501;
            proxy_http_version 1.1;
            proxy_set_header Upgrade $http_upgrade;
            proxy_set_header Connection "upgrade";
            proxy_set_header Host $host;
            proxy_set_header X-Real-IP $remote_addr;
            proxy_set_header X-Forwarded-For $proxy_add_x_
            forwarded_for;
            proxy_set_header X-Forwarded-Proto $scheme;
        }

        # location /_stcore/stream {
        #     include proxy_params;
        #     proxy_pass http://streamlit:8501/_stcore/stream;
        # }
    }
server {
    listen 8504 ssl;
    server_name lb-webserver.pro www.lb-webserver.pro;
    root           /usr/share/nginx/html;

    ssl_certificate "/etc/pki/nginx/lb-webserver_pro_
    combined.crt";
    ssl_certificate_key "/etc/pki/nginx/private/
    private.key";
    ssl_session_cache shared:SSL:1m;
    ssl_session_timeout  10m;

    include /etc/nginx/default.d/*.conf;

    error_page 404 /404.html;
```

```
    location = /40x.html {
    }

    error_page 500 502 503 504 /50x.html;
    location = /50x.html {
    }

location / {
     include proxy_params;
     proxy_pass http://jenkins:8080;  # Make sure this
     matches the internal port Jenkins is running on
    }
  }
}
```

4. **Access Jenkins**

    - Jenkins should be now up and running through port 8504. Open
      a browser and navigate to `https://<your-domain>:8504/` (e.g.,
      `https://lb-webserver.pro:8504/`). You will be prompted to
      enter an initial admin password (see Figure 5-1).

    - Retrieve the password from the file `jenkins/var/jenkins_home/`
      `secrets/initialAdminPassword` in the repository folder, paste it
      in the browser, and click continue.

    - Next, select the option to install the suggested plug-ins (see
      Figure 5-2), and wait for the installation to complete.

    - Once finished, create the first admin user, then click **Save and
      Continue** in the bottom-right corner.

    - When prompted, add the requested URL (e.g., `https://lb-`
      `webserver.pro:8504/`), and confirm.

    - After seeing the message **"Jenkins is ready!"**, click **Start using
      Jenkins** to finalize the setup. You're all set!

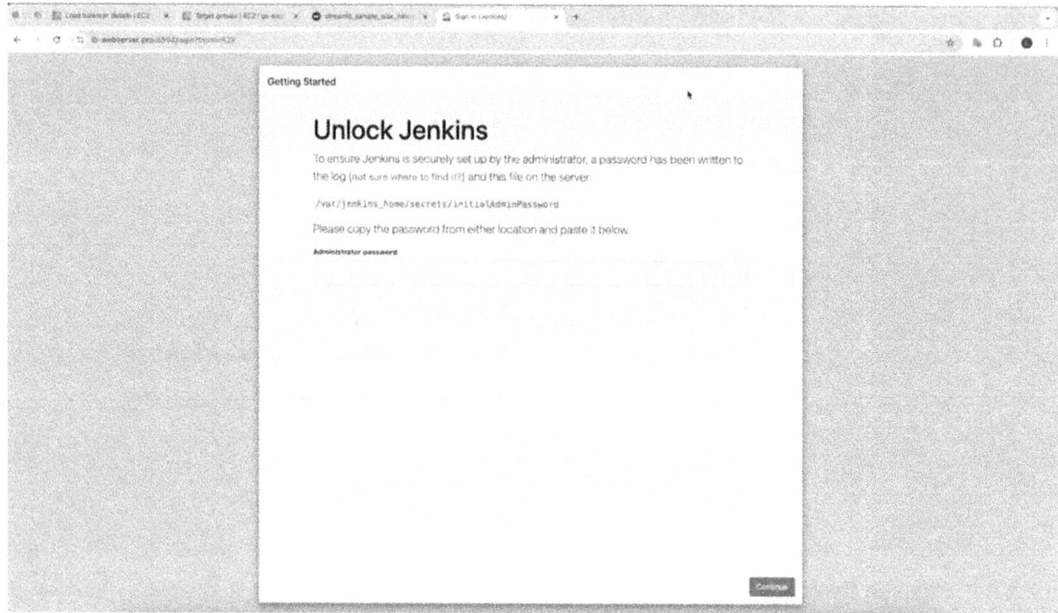

Figure 5-1. *Unlock Jenkins*

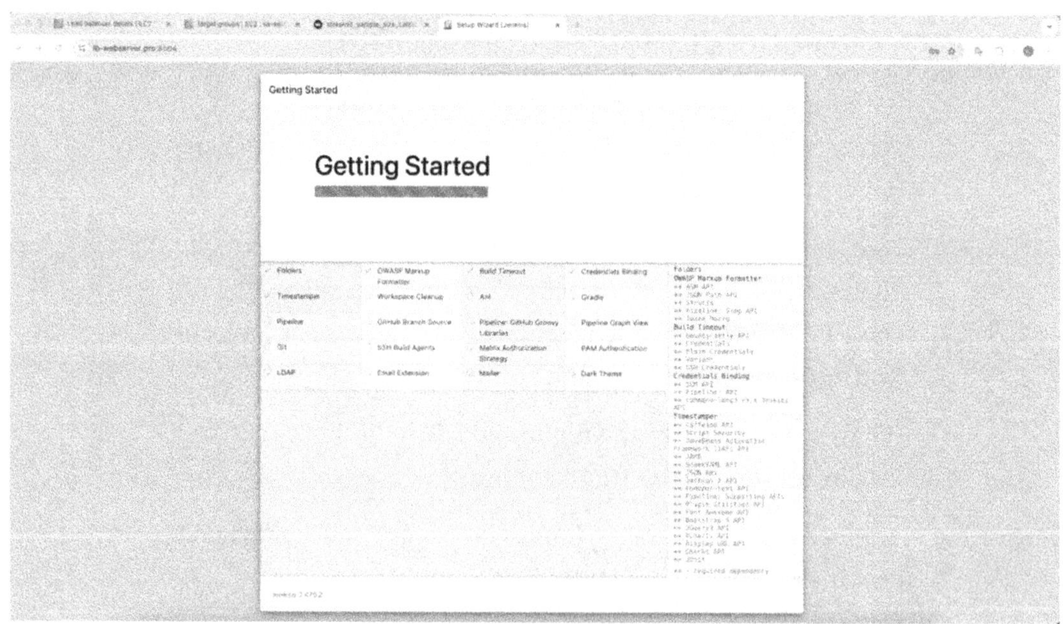

Figure 5-2. *Installing plug-ins*

Enable Docker Service to Start Automatically

After deploying the final application with Docker, if the EC2 instance is turned off and then restarted, the application will not automatically start. You will need to reconnect via SSH and manually rerun Docker to get it running again. To avoid this, you can configure Docker to restart automatically whenever the instance is powered back on by using the commands in Code-block 5-4 and shown in Figure 5-3.

This is particularly useful in scenarios where the EC2 instance is scheduled to shut down during non-working hours and restart during working hours. Enabling this service eliminates the need to SSH into the instance each time, ensuring the application starts automatically.

Code-block 5-4. Enable Docker service to start up automatically

```
# Enable the Service
sudo systemctl enable docker

# Verify the Configuration
sudo systemctl status docker
```

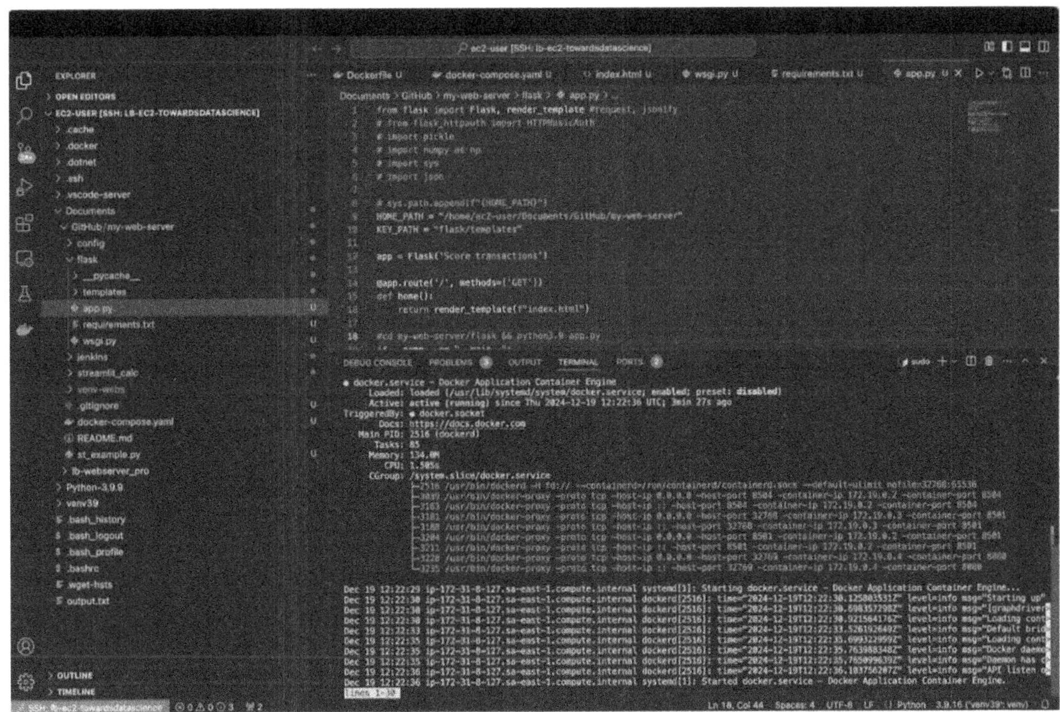

Figure 5-3. *Enable Docker service to start up automatically*

Train a Model (Prepare VS Code for Jupyter Notebooks)

Before deploying the Flask API, we will train a machine learning model using Jupyter notebooks in VS Code. This section ensures VS Code is configured to work seamlessly with notebooks.

Steps to Configure VS Code for Notebooks

Before training a model in VS Code, you need to set it up to work with Jupyter notebooks (files with the `.ipynb` extension).

1. **Open the notebook**

   - In VS Code, open the `model_training.ipynb` file located in the `notebooks` folder of the repository.

   - Click **Select Kernel** in the top-right corner. A pop-up will appear in the middle of the screen. Click **Install/Enable suggested extensions**.

2. **Install necessary extensions**

   - After enabling the suggested extensions, click **Select Kernel** again and choose **Browse marketplace for kernel extensions**.

   - Install the **Jupyter extension** by Microsoft (make sure to select **Install in SSH**).

   - Next, go to the **Extensions** panel in VS Code (left-hand side) and search for the **Python extension** by Microsoft. Ensure it is installed.

3. **Select the virtual environment**

   - Click **Select Kernel** once more, and choose **Python Environments....** From the list, select the virtual environment you created in the repository folder (e.g., `venv-webs`).

   - If the virtual environment does not appear, restart VS Code and try again.

4. **Alternative method**

- If the above steps don't work, use the search bar in VS Code (or press **Command+Shift+P** on macOS) to search for **Select a Python Environment**. Select your virtual environment from the list.

- If a pop-up prompts you to install additional packages like `ipykernel`, follow the instructions to complete the installation. You may need to restart VS Code again after this step.

5. **Install additional packages**

- Ensure the kernel used to train the model matches the virtual environment you created. This allows you to install packages specifically for this project.

- Open the terminal, activate the virtual environment, and install required packages like `scikit-learn` using the `pip` command. Refer to Code-block 5-5 for the exact command.

By ensuring that your virtual environment is properly set up and linked to the Jupyter notebook, you'll be able to train models seamlessly and install any required dependencies for your specific project.

Code-block 5-5. pip install scikit-learn package

```
cd ~/Documents/GitHub/deploy-secure-ds-apps-book
source venv-webs/bin/activate
pip install scikit-learn
```

Train the Model

1. **Import libraries**

- To begin training your model, start by importing the necessary libraries, see Code-block 5-6. For machine learning tasks, use `sklearn`, which provides access to a wide range of well-known ML models as well as utility functions for splitting data into training and validation sets and for calculating metrics such as accuracy, precision, and recall. Additionally, import `pandas` for working with dataframes and `numpy`, a powerful library for matrix manipulation and numerical computations.

Code-block 5-6. Import libraries

```python
import pandas as pd
import numpy as np
from sklearn.datasets import make_classification
from sklearn.model_selection import train_test_split
from sklearn.ensemble import RandomForestClassifier
from sklearn.metrics import accuracy_score, precision_score,
recall_score, classification_report
import pickle
```

2. **Load or create the data**

- In this example, we are generating a synthetic dataset using the
  make_classification function from sklearn; see Code-block 5-7.
  The dataset consists of 1,000 rows, four predictor features, and one
  target feature. All four predictors are informative, and the target
  feature represents a binary classification. The dataset is balanced,
  with equal representation of both classes.

- To visualize the dataset created, we run the code from
  Code-block 5-8. The dataset is shown in Figure 5-4.

Code-block 5-7. Load data

```python
X, y = make_classification(
    n_samples=1000,        # Total samples
    n_features=4,          # Number of predictors
    n_informative=4,       # All features are informative
    n_redundant=0,         # No redundant features
    n_classes=2,           # Binary classification
    weights=[0.5, 0.5],    # Balanced dataset
    random_state=42        # Reproducibility
)
data = pd.DataFrame(X, columns=[f'Feature_{i}' for i in
range(1, 5)])
data['Target'] = y
```

Code-block 5-8. See the data shape and visualize the dataframe

```
# print the number of rows and columns of the dataset
print(data.shape)
# visualize the dataset
data.head()
```

In [6]:   data

Out[6]:

	Feature_1	Feature_2	Feature_3	Feature_4	Target
0	-2.439926	-1.371106	-0.025787	-0.102379	0
1	-0.366951	1.380114	0.583490	0.443968	1
2	-1.321133	-0.680427	0.749495	-0.281137	1
3	-0.330055	1.443556	0.508346	1.019528	1
4	-1.211829	-2.021848	0.143790	0.501954	0
...	...	...	...	...	...
995	-1.565182	-0.913232	0.495010	0.621762	1
996	-1.600971	2.692596	0.311441	2.504642	1
997	0.043325	-0.200804	-1.827989	-1.952151	0
998	0.259675	1.685565	-0.112371	1.995352	1
999	-3.193412	-1.021301	0.041412	0.140774	0

1000 rows × 5 columns

Figure 5-4. *Visualize data*

3. **Split the dataset**

- Next, split the dataset into training and validation sets. This step ensures that your model is trained on one portion of the data and validated on another to evaluate its performance. In this example, the dataset is randomly split, with 80% of the rows used for training and 20% reserved for validation. See Code-block 5-9.

Code-block 5-9. Split the dataset into train and test

```
X_train, X_test, y_train, y_test = train_test_split(
    data.drop(columns='Target'),
    data['Target'],
```

```
    test_size=0.2,
    random_state=42
)
```

4. **Model training**

    - Train a Random Forest model using the training dataset,
      which contains 80% of the samples. In this example, we use the
      RandomForestClassifier from sklearn with default parameters.
      The model is then fitted to the training data to learn the relationships
      between the features and the target variable. See Code-block 5-10.

    Code-block 5-10. Train the model

    ```
    clf = RandomForestClassifier(random_state=42)
    clf.fit(X_train, y_train)
    ```

5. **Evaluate the model**

    - Since the trained model is a binary classification model
      (predicting a target of 0 or 1), we will evaluate its performance
      using metrics such as precision, recall, and F1-score. These
      metrics provide a comprehensive understanding of the model's
      ability to make accurate predictions. The evaluation is performed
      on the validation set, which consists of unseen data to ensure an
      unbiased assessment. See Code-block 5-11 for the commands
      and visualize the classification report shown in Figure 5-5.

    Code-block 5-11. Evaluate the model

    ```
    y_pred = clf.predict(X_test)

    accuracy = accuracy_score(y_test, y_pred)
    precision = precision_score(y_test, y_pred)
    recall = recall_score(y_test, y_pred)
    report = classification_report(y_test, y_pred)
    print(f"Accuracy: {accuracy:.2f}")
    print(f"Precision: {precision:.2f}")
    print(f"Recall: {recall:.2f}")
    print("\nClassification Report:\n", report)
    ```

```
Accuracy: 0.90
Precision: 0.92
Recall: 0.89

Classification Report:
              precision    recall  f1-score   support

           0       0.88      0.92      0.90        95
           1       0.92      0.89      0.90       105

    accuracy                           0.90       200
   macro avg       0.90      0.90      0.90       200
weighted avg       0.90      0.90      0.90       200
```

Figure 5-5. *Model evaluation results*

6. **Save the model object**

- To persist the trained model, we will use `pickle` to save the model object to a file. This allows us to reload the model later for predictions or further analysis.

- The first part of the code saves the `RandomForestClassifier` model (`clf`) to the specified file path. The second part demonstrates how to reload the saved model for later use. By saving the model, you ensure that it can be reused without retraining, saving time and computational resources. See Code-block 5-12.

Code-block 5-12. Save the model object in pickle format

```python
HOME_PATH = "/home/ec2-user/Documents/GitHub/deploy-secure-ds-apps-book"
# Save the model object (clf) to a file
with open(f'{HOME_PATH}/flask/models/random_forest_model.pkl',
'wb') as model_file:
    pickle.dump(clf, model_file)

# Load the saved model object when needed
with open(f'{HOME_PATH}/flask/models/random_forest_model.pkl',
'rb') as model_file:
    model_obj = pickle.load(model_file)
```

7. **Create a payload and predict**

- This step replicates the functionality your Flask API will perform. The API will receive a POST request with a payload in a format similar to the one shown in Code-block 5-11, process it, and return a prediction score between 0 and 1. The prediction score represents the model's confidence in classifying the input as one of the target classes. For example, a score of 0.65 indicates the model leans toward class 1; a score of 0.85 reflects an even stronger preference for class 1.

- By default, any score above 0.5 is classified as target=1, while scores below 0.5 are classified as target=0. However, you can adjust this threshold by running a threshold optimization loop on the validation set to determine the best threshold for your specific use case. See Code-block 5-13 showing how to arrange the data in the format of a payload.

Code-block 5-13. Create a payload

```
data = {
    "Feature_1": 1.0,
    "Feature_2": 1.0,
    "Feature_3": 1.0,
    "Feature_4": 1.0
}
data.values()
data_np = np.array(list(data.values())).reshape(1, -1)
# array([[1., 1., 1., 1.]])
```

- Once the payload is prepared, the model processes it to generate a prediction score. The example below demonstrates how the payload is transformed into a prediction score, as shown in Code-block 5-14.

Code-block 5-14. Predict the score using the payload

```
y_proba = model_obj.predict_proba(data_np)
predicted_class = np.argmax(y_proba[0])
```

```
predicted_score = y_proba[0][predicted_class]
print(f'predicted score {predicted_score}')
# predicted score 0.99
```

- The `predict_proba` method calculates the probabilities for each class, and the `argmax` function identifies the class with the highest probability. The `predicted_score` variable then stores the confidence score for the predicted class, which can be returned by the API for further decision-making.

Note   In the context of an API, the **request** refers to the entire HTTP request sent to the server, including the method (e.g., GET, POST), headers, and other metadata. The **payload**, on the other hand, is the body of the request, which contains the actual data being sent to the API. For example, in this chapter, the `data` dictionary represents the payload that will be sent as part of the request.

Deploy a Flask App for API Calls with Authentication

In this section, we will deploy a Flask application that serves API calls with authentication. The Flask app's primary function is to process incoming payloads, use the trained model to make predictions, and return the results to the requester. Before deploying the application with Docker, we will test it locally.

Steps to Deploy the Flask App

1. **Clone the Flask app code**

   - The complete code for the Flask app is available at book's GitHub repository under the respective chapter.

2. **Test locally**

   - To ensure the Flask app is working as expected, you can test it locally before deploying it. Navigate to the `flask` folder and run the `app.py` file from the terminal. Use the commands from Code-block 5-15.

Code-block 5-15. pip install scikit-learn package

```
cd ~/Documents/GitHub/deploy-secure-ds-apps-book/flask
source venv-webs/bin/activate
python app.py
```

- The application will launch on localhost:8502. You can test its functionality following the next few code-blocks or by using the provided notebook 'post_predict.ipynb' available at book's GitHub repository under the respective chapter.

i. **Import libraries and define functions**

- In this step, we import the necessary libraries and define utility functions to handle HTTP requests and payloads. The requests library is used for making GET and POST requests, while numpy (previously introduced) is used for data manipulation. Additionally, the json library is included for managing payloads, and the HTTPBasicAuth class from requests.auth is available for authentication purposes. See Code-block 5-16 for the imports and function definitions.

Code-block 5-16. Import libraries and functions

```python
import requests
import numpy as np
from requests.auth import HTTPBasicAuth
import json

def get_response(url, timeout=False):
    print(url)
    if timeout:
        response = requests.get(url, timeout=timeout)
        print(response.status_code)
    else:
        response = requests.get(url)
    print(response)
```

```python
def post_response(url, json_data):
    print(f"Sending POST request to {url} with data:
    {json_data}")
    response = requests.post(url, json=json_data)
    print(f"Response status code: {response.status_code}")
    if response.status_code == 200:
        print(f"Response JSON: {response.json()}")
        return response.json()
    else:
        print(f"Error: {response.text}")
        return None
```

ii. **Send a GET request**

- Using the get_response() function, we send a GET request
  to the API's endpoint. The function takes the URL as an input
  parameter and returns a response. If the response code is
  200, it indicates that the request was successful and the API is
  functioning correctly.

  See Code-block 5-17.

  Code-block 5-17. Send a GET request

```python
get_response(url = "http://localhost:8502/")
# http://0.0.0.0:8502/
# <Response [200]>
```

ii. **Create a payload and send a POST request**

- In Code-block 5-18, we create a payload similar to the one
  in Code-block 5-13. The payload includes the input features
  required by the API. Using the post_response() function,
  which takes the URL and the payload as input parameters,
  we send a POST request to the API's /predict endpoint. This
  request returns the response status and a prediction score for
  the provided payload.

Code-block 5-18. Send a POST request

```
data = {
    "Feature_1": 2.0,
    "Feature_2": 2.0,
    "Feature_3": 2.0,
    "Feature_4": 2.0
}

response = post_response(url = "http://localhost:8502/
predict", json_data=data)

# Sending POST request to http://localhost:8502/predict
#      with data: {'Feature_1': 2.0, 'Feature_2': 2.0,
'Feature_3': 2.0, 'Feature_4': 2.0}
# Response status code: 200
# Response JSON: {'model_name': 'random_forest_model',
'score': 0.77}
```

Note   Understanding HTTP Response Codes

HTTP response codes indicate the outcome of a request made to an API. Here are some common response codes:

- **200 (OK)**: The request was successful, and the server returned the expected result.

- **201 (Created)**: The request was successful, and a resource was created (typically for POST requests).

- **400 (Bad Request)**: The server could not understand the request due to invalid syntax or missing data.

- **401 (Unauthorized)**: Authentication is required, and the provided credentials are invalid or missing.

- **403 (Forbidden)**: The request is understood, but the server is refusing to fulfill it (e.g., insufficient permissions).

- **404 (Not Found)**: The requested resource or endpoint could not be found.

- **500 (Internal Server Error)**: The server encountered an unexpected condition that prevented it from fulfilling the request.

These codes help debug issues with API communication and determine if additional steps, like correcting the request or updating authentication, are needed.

---

3. **Prepare the Flask app for deployment**

- Before deploying the Flask app with Docker, create the following folders and files within the `flask` directory:

   i. **Create the `logs` folder**

   - This folder is used as a volume in the `docker-compose.yaml` file. If you do not want to create this folder, comment out the corresponding volume line in the `docker-compose.yaml` file.

   ii. **Create the `config/auth_flask.json` file**

   - This file stores authentication credentials for the Flask app. Use the format shown in Code-block 5-19.

      Code-block 5-19. auth_flask.json format

```
{
    "prod": {
        "user": "admin",
        "password": "password"
    }
}
```

   iii. **Grant permissions to the folders**

   - Similar to the steps in Code-block 5-20, grant appropriate permissions to the newly created folders and files so Docker has access to them. Use the following commands:

Code-block 5-20. Authorizing recently created folder and files

```
cd ~/Documents/GitHub/deploy-secure-ds-apps-book/flask
sudo chown -R :jenkins_shared
chmod 775 -R logs/
chmod 775 -R config/
```

4. Deploy the Flask app with docker-compose

   - Once the setup is complete, navigate to the repository's main folder and deploy all applications, including the Flask app, using Docker Compose; see Code-block 5-21.

   Code-block 5-21. Run docker-compose

   ```
   cd ~/Documents/GitHub/deploy-secure-ds-apps-book
   docker-compose up -d –build
   ```

   - The Flask app will now be deployed along with the other applications (Jenkins and Streamlit).

5. **Verify the deployment**

   - After deploying thc application, you can verify the Flask app's functionality by accessing the web page it generates. The Flask app will create a page, similar to Figure 5-6, showing detailed instructions on how to use the API for predictions. The specific URL will depend on the domain you configured during deployment. By using the provided code in the GitHub repository, the page will be adapted to your setup.

     i. For example, `https://lb-webserver.pro:8502/`

        - Additionally, you can verify the deployment of the other applications, such as the Streamlit dashboard and Jenkins server. These, too, will be accessible via the domain you configured. Replace the example URLs below with your specific domain and ports:

   i.   Flask app (API instructions): `https://<your-domain>:8502/`

   ii.   Streamlit dashboard: `https://<your-domain>:8501/`

   iii.   Jenkins server: `https://<your-domain>:8504/`

6. **Test the Flask app**

- To test the Flask app, you have several options:

   i.   **Using the provided notebook**

  - Navigate to the HTTPS section of the notebook 'post_predict.ipynb' provided at book's GitHub repository under the respective chapter.

   ii.   **Using the steps below**

  - Follow the steps outlined here to interact with the API. Begin by importing the necessary libraries and defining functions from Code-block 5-16. You can reuse the payload provided in Code-block 5-18 to send requests.

Send an Authenticated Request

To send an authenticated request to the deployed application, start by importing the authentication keys from the `auth_flask.json` file. This file contains the credentials needed to securely interact with the API; see Code-block 5-22.

Code-block 5-22. Run docker-compose

```python
PATH_AUTH = '/Users/l.benevides/Documents/personal/aws/auth_flask.json'

with open(f'{PATH_AUTH}', 'r') as file:
    json_api_key = json.load(file)
```

Send an Authenticated GET Request

Use Code-block 5-23 to send an authenticated GET request to the safe URL deployed in the production environment with Docker.

Code-block 5-23. Authenticated GET request

```python
url = "https://lb-webserver.pro:8502/"

response = requests.get(url, auth=HTTPBasicAuth(json_api_key['prod']
['user'], json_api_key['prod']['password']))
print(response)
```

Send an Authenticated POST Request

Next, use Code-block 5-24 to send an authenticated POST request. Note that this uses a different payload, so the predicted score will differ.

Code-block 5-24. Authenticated POST request

```python
data = {
    "Feature_1": 1.0,
    "Feature_2": 1.0,
    "Feature_3": 1.0,
    "Feature_4": 1.0
}

response = requests.post(
    url = "https://lb-webserver.pro:8502/predict",
    json=data,
    auth=HTTPBasicAuth(json_api_key['prod']['user'], json_api_key['prod']
    ['password'])
)
print(response.text)
# {"model_name":"random_forest_model","score":0.99}
```

By following these steps, you can verify that your deployed Flask application is functioning as expected. The GET request confirms access to the API, while the POST request returns predictions from the model trained earlier in the section "Train a Model (Prepare VS Code for Jupyter Notebooks)."

Figure 5-6. *Flask index html*

The Flask app uses the model trained in the section "Train a Model (First Prepare VS Code to run .ipynb)" to process incoming API requests and return predictions. For a visual demonstration, see Figure 5-6, which shows the Flask app's web page and API instructions.

Summary

In this chapter, you have deployed and tested a secure and production-grade deployment architecture including a server for ETL pipelines and machine learning model serving API with authentication. More specifically you have

- Deployed Jenkins on port 8504 to manage ETL pipelines, ensuring data persistence via Docker volumes

- Enabled Docker to start automatically with system reboot using systemctl, ensuring application uptime

- Configured VS Code for Jupyter notebooks to support model development using .ipynb files remotely on your EC2 instance

- Trained and evaluated a Random Forest model using synthetic classification data with scikit-learn and saved the model using pickle for reuse in production

  - Scikit-learn documentation: `https://scikit-learn.org/stable/`

- Built and tested a Flask API with authentication that receives payloads, applies the model, and returns prediction scores

  - Flask documentation: `https://flask.palletsprojects.com/en/stable/`

- Deployed Flask, Jenkins, and Streamlit apps with Docker Compose and made them accessible over HTTPS using your registered domain

  - Docker Compose documentation: `https://docs.docker.com/compose/`

  - NGINX Reverse Proxy documentation: `https://docs.nginx.com/nginx/admin-guide/web-server/reverse-proxy/`

In the next chapter, you'll build on this multi-app deployment by organizing your infrastructure using subdomains for each application. You'll learn how to adjust Docker Compose and Nginx accordingly so that it routes traffic to each application. You'll secure all subdomains with SSL certificates and reorganize AWS target groups and security groups to align with the new subdomain architecture; at the end you'll deploy and validate the multi-subdomain production environment.

## CHAPTER 6

Create and Secure Your Subdomains

In the previous chapter, you deployed a production-ready environment using Docker and Nginx to host Jenkins, Flask, and Streamlit applications. You configured authentication for Flask, enabled Docker services to persist on reboot, and trained a machine learning model that could be served through your Flask API.

In this chapter, we'll walk through the process of setting up and securing subdomains for each one of your applications (e.g., "flask.lb-webserver.pro", "streamlit.lb-webserver. pro", "jenkins.lb-webserver.pro"). First, we'll modify the Nginx configuration file (`nginx. conf`) and update the Docker Compose YAML file (`docker-compose.yaml`) to define subdomains for your applications. Additionally, we'll make the necessary changes to the **AWS security groups** and **load balancer target groups** to allow traffic to the subdomains. Once that's done, we'll deploy the applications. However, at this stage, your subdomains will only work over HTTP, which means they won't be secure just yet.

To fully secure these subdomains, you'll also generate and apply a Multi-domain SSL Certificate using Namecheap and OpenSSL. By the end of this chapter, you'll have a secure setup with HTTPS for all your subdomains, ensuring your infrastructure meets professional standards and protects your applications from vulnerabilities.

Set Up Nginx and Docker-Compose

A **subdomain** is an extension of your primary domain that lets you host multiple applications or services under the same domain. For example, you could have `api. example.com` for your API, `dashboard.example.com` for your dashboards, and `jenkins.`

© Lucas H. Benevides e Braga 2025
L.H.B.e. Braga, *Deploying Secure Data Science Applications in the Cloud*,
https://doi.org/10.1007/979-8-8688-1715-1_6

example.com for automation tools, all linked to example.com. Subdomains are not only helpful for organizing your infrastructure but also for separating different applications and making them easier to access and manage.

Nginx Configuration Changes

To configure subdomains for your applications, the first step is to update the reverse proxy, **Nginx**, by editing the nginx.conf file. Instead of having each server listen directly on its corresponding port (e.g., Streamlit on 8501, Flask on 8502, and Jenkins on 8504), we'll configure them to listen on the HTTPS port 443. Each application will be redirected to its specific subdomain. Below are the key changes to be made in the nginx.conf file. See Table 6-1 for a before-and-after comparison.

Table 6-1. *Before and after changes on nginx.conf*

Before	After
Each application listens to its assigned port on the primary domain	Each application is now redirected to a subdomain and listens on port 443 (HTTPS)

```
# Streamlit               # Streamlit
server {                  server {
  listen 8501 ssl;          listen 443 ssl;
  server_name lb-webserver.pro;   server_name streamlit.lb-webserver.pro;
  [...]                     [...]
  location / {             location / {
    include proxy_params;    include proxy_params;
    proxy_pass http://streamlit_   proxy_pass http://streamlit_
    calc:8501;               calc:8501;
    [...]                    [...]
  }                        }
}                        }
```

(*continued*)

Table 6-1. (*continued*)

Before	After
Each application listens to its assigned port on the primary domain	Each application is now redirected to a subdomain and listens on port 443 (HTTPS)

```
# Flask
server {
  listen 8502 ssl;
  server_name lb-webserver.pro;
  [...]
  location / {
    include proxy_params;
    proxy_pass
    http://flask:8502;
    [...]
  }
}
# Jenkins
server {
  listen 8504 ssl;
  server_name lb-webserver.pro;
  [...]
  location / {
    include proxy_params;
    proxy_pass
    http://jenkins:8080;
    [...]
  }
}
```

```
# Flask
server {
  listen 443 ssl;
  server_name flask.lb-webserver.pro;
  [...]
  location / {
    include proxy_params;
    proxy_pass http://flask:8502;
    [...]
  }
}
# Jenkins
server {
  listen 443 ssl;
  server_name jenkins.lb-webserver.pro;
  [...]
  location / {
    include proxy_params;
    proxy_pass http://jenkins:8080;
    [...]
  }
}
```

Docker Compose Configuration Changes

Next, we need to update the `docker-compose.yaml` file to synchronize it with the changes made in the `nginx.conf` file. Specifically, we'll adjust the ports exposed by Nginx. Instead of using ports 8501, 8502, and 8504, Nginx will now expose the standard HTTP (80) and HTTPS (443) ports; see Table 6-2.

Table 6-2. *Before and after changes on docker-compose.yaml*

Before	After
`[...]`	`[...]`
`nginx:`	`nginx:`
`[...]`	`[...]`
`depends_on:`	`depends_on:`
`- streamlit_calc`	`- streamlit_calc`
`- jenkins`	`- jenkins`
`- flask`	`- flask`
`ports:`	`ports:`
`- "8501:8501"`	`- "80:80"`
`- "8504:8504"`	`- "443:443"`
`- "8502:8502"`	

The full updated configuration files are available in the GitHub repository:

- nginx.conf

- docker-compose.yaml

Note   Understanding Ports 80 and 443

Ports 80 and 443 are standard ports for web traffic:

- **Port 80:** Used for HTTP (unencrypted) traffic

- **Port 443:** Used for HTTPS (encrypted) traffic

These ports are globally recognized and allow browsers to automatically connect without requiring them in the URL suffix. For example, when accessing `https://streamlit.lb-webserver.pro`, the browser assumes port 443 is being used, and you don't need to specify it explicitly.

By making these updates, your applications will be accessible through their respective subdomains over HTTPS, providing a more professional and secure infrastructure.

Set Up Security Groups and Target Groups on AWS

This process is similar to what we covered in Chapter 3, where we created security groups and target groups for the load balancer. In this case, we'll be modifying the existing configurations to accommodate ports **443** (for HTTPS) and **80** (for HTTP). These changes will apply to the security groups of both the load balancer and the EC2 instance, as well as to the load balancer's listeners that connect to the target groups.

Additionally, as we no longer need ports **8501**, **8502**, and **8504**, it's a good practice to close these ports in all places where they were previously open. This includes the security groups and target groups. Removing unnecessary open ports helps minimize potential security vulnerabilities.

Step 1: Update the Security Group for the Load Balancer

Start by modifying the load balancer's security group to reflect the new configuration. See Figure 6-1 for the final state of the load balancer's security group.

- **Port 22:** This remains open for SSH connections.

- **Ports 8501, 8502, and 8504:** Remove these ports since they're no longer required.

- **Ports 80 and 443:** Add these ports to enable HTTP and HTTPS traffic.

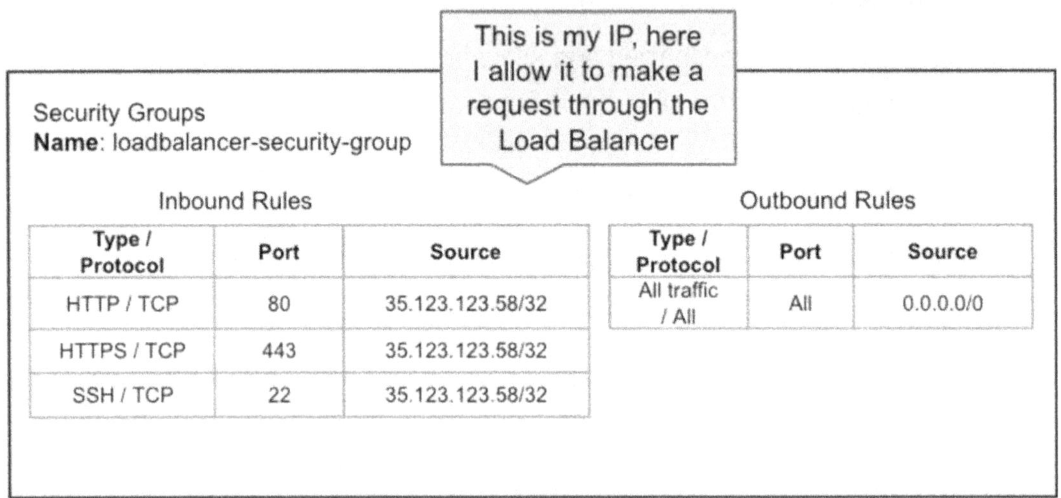

Type / Protocol	Port	Source
HTTP / TCP	80	35.123.123.58/32
HTTPS / TCP	443	35.123.123.58/32
SSH / TCP	22	35.123.123.58/32

Type / Protocol	Port	Source
All traffic / All	All	0.0.0.0/0

Figure 6-1. *Final state of the load balancer's security group*

Step 2: Update the Security Group for the EC2 Instance

Similarly, update the EC2 instance's security group. Refer to Figure 6-2 for the final configuration. Follow the same process:

- **Remove ports 8501, 8502, and 8504.**

- **Add ports 80 and 443** to handle HTTP and HTTPS traffic.

If you need guidance on how to add or remove inbound rules in a security group, refer back to the step-by-step instructions in Chapter 3.

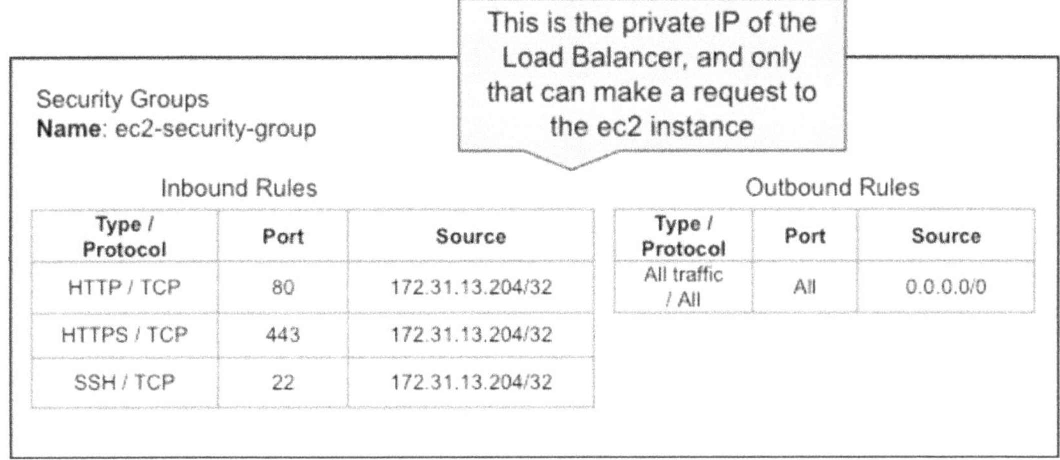

Figure 6-2. *Final state of the EC2's security group*

Step 3: Update the Load Balancer's Listeners and Target Groups

Next, configure the load balancer to listen on ports **443** and **80** and update the target groups to match. See Figure 6-3 for the final state of the target groups.

- **Add listeners for ports 80 and 443:** These will handle HTTP and HTTPS traffic and forward requests to the target groups.

- **Remove any listeners for ports 8501, 8502, and 8504:** These ports are no longer used in the updated configuration.

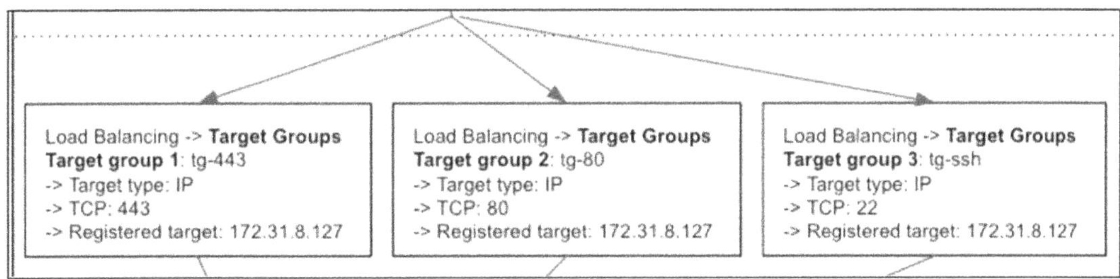

Figure 6-3. *Final state of the target groups*

Final State of the Infrastructure

After completing these steps, your infrastructure is now set up to support subdomains using ports **80** and **443**, and the traffic is routed securely via the load balancer. You can now run docker-compose to deploy everything you've developed in the last two sections. Once the deployment is complete, you'll be able to access your applications using the subdomains (flask.lb-webserver.pro, streamlit.lb-webserver.pro, jenkins.lb-webserver.pro). However, when you open these URLs, you'll notice that the **HTTPS** in the browser's address bar has a strikethrough and is highlighted in red. This indicates that the connection is not secure. This issue occurs for two reasons:

1. **Inappropriate SSL certificate:** While you have an SSL certificate, the one currently in use is not suitable for securing subdomains.

2. **CSR configuration:** During the certificate signing request (CSR) process in Chapter 3, subdomains were not explicitly included in the request for the SSL certificate.

To address these issues and fully secure your subdomains, proceed to the section "Multi-subdomain with SSL Certificate," where we'll guide you through purchasing an SSL certificate that supports subdomains and updating your infrastructure to ensure all traffic is securely encrypted. See Figure 6-4 for an overview of the updated infrastructure with subdomains.

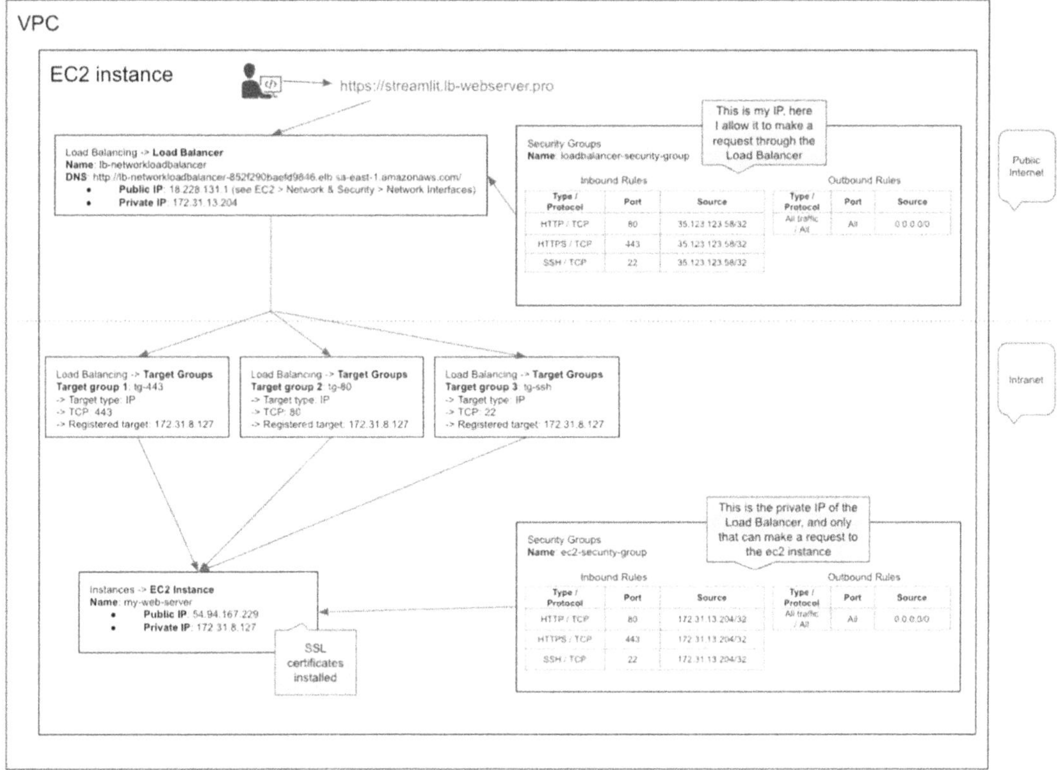

Figure 6-4. *Final state of the infrastructure with subdomains*

Multi-subdomain with SSL Certificate

Choosing the Right Certificate

To secure your subdomains with HTTPS, you'll need to purchase an SSL certificate that supports subdomains. There are several types of SSL certificates available, and in this section, we'll use namecheap.com as an example for purchasing both the domain and certificates. Two common options for subdomain security are **Wildcard SSL Certificates** and **Multi-domain SSL Certificates**. Let's break them down and examine their pros, cons, and costs.

- **Wildcard SSL Certificates:** Wildcard certificates provide significant flexibility by automatically covering the primary domain and any number of subdomains under it. For example, a wildcard certificate for `example.com` would secure `api.example.com`, `dashboard.example.com`, and any future subdomains without needing to reissue the certificate.

- **Pros**

  - Unlimited subdomains are covered under the primary domain.

  - Flexibility to add new subdomains at any time without needing to reissue the certificate.

- **Cons**

  - Typically more expensive than multi-domain certificates. On namecheap.com, a wildcard SSL certificate costs **$89.99 for the first year** (if purchased for one year).

- **Multi-domain SSL Certificates:** Multi-domain certificates, also called SAN (Subject Alternative Name) certificates, allow you to secure multiple specific domains or subdomains. For instance, you can secure `example.com`, `api.example.com`, and `dashboard.example.com`. However, this option comes with some limitations:

  - You need to specify all the subdomains at the time of issuing the certificate.

  - If you need to add more subdomains later, you must reissue the certificate.

  - There is a limit on the number of domains and subdomains (SANs) included, and additional SANs come at an extra cost.

  - **Example**

    - A typical multi-domain SSL certificate includes **three SANs** by default. In our case, we need to secure

      1. Primary domain: `lb-webserver.pro`

      2. Subdomains: `streamlit.lb-webserver.pro`, `flask.lb-webserver.pro`, and `jenkins.lb-webserver.pro`

- Since we have **four SANs in total**, we would need to purchase an additional SAN for around **$21.59**. The base cost for a multi-domain SSL certificate on namecheap.com is **$35.99 for the first year**, so the total cost would be **$35.99 (certificate) + $21.59 (additional SAN) = $57.58**.

- **Pros**

  - Typically less expensive than wildcard certificates

  - Suitable for cases where the number of domains and subdomains is fixed and not likely to change

- **Cons**

  - Limited to a fixed number of SANs (you must pay extra for additional SANs)

  - Requires reissuing the certificate if you need to add new subdomains later

Which Option Is Best for This Use Case?

For the purpose of this book, a **Multi-domain SSL Certificate** is the better choice, as we are only securing three applications (`streamlit`, `flask`, and `jenkins`) under `lb-webserver.pro`. Since these applications are fixed and unlikely to change, the cost of a multi-domain certificate is more economical.

However, if you anticipate needing additional subdomains in your use case (e.g., a new subdomain for another Streamlit dashboard or application), the costs for additional SANs can quickly add up. For example, securing five SANs would bring the total cost of a multi-domain certificate close to that of a wildcard certificate. In such cases, a **Wildcard SSL Certificate** might be the better investment, as it provides unlimited subdomain flexibility for a fixed price. Multiple options are shown in Figure 6-5 from namecheap.com.

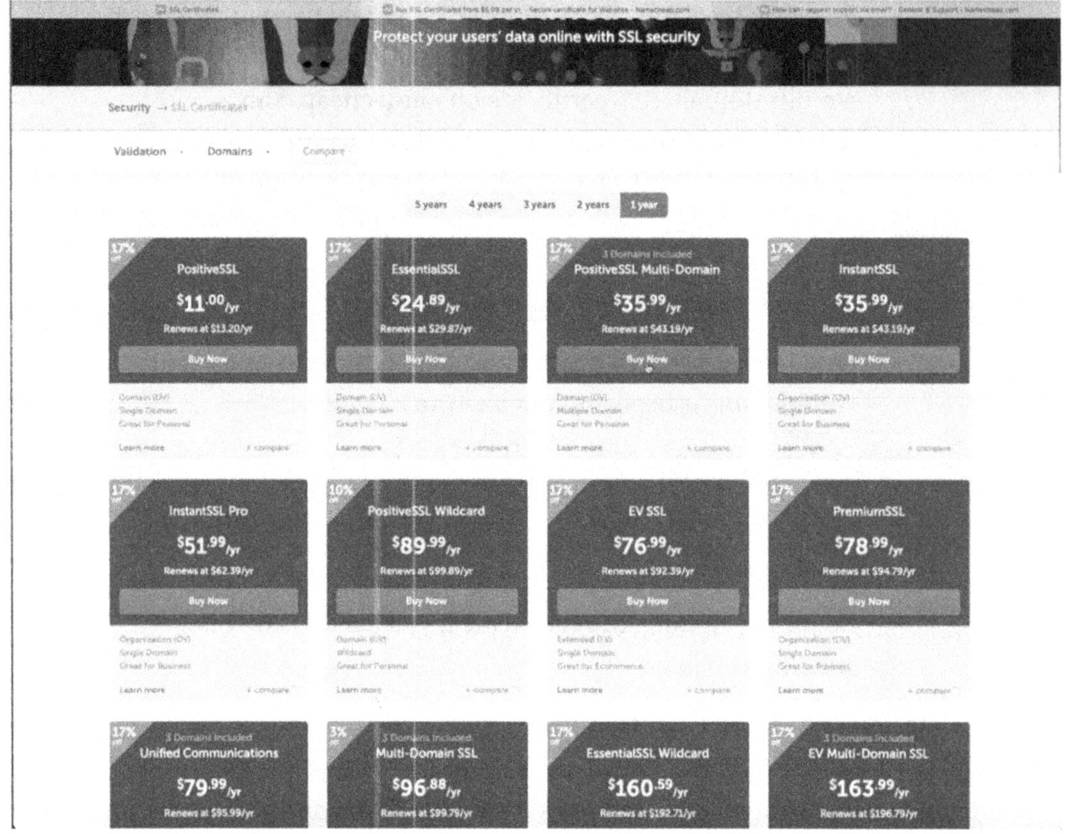

Figure 6-5. *Comparison of SSL certificate options and pricing on namecheap.com*

Certificate Signing Request (CSR)

Now that we've discussed the different SSL certificate options and decided to use the **Multi-domain SSL Certificate**, let's move forward with the installation. The process will be similar to what we did in Chapter 4, where we created a certificate signing request (CSR). However, for this multi-domain setup, there are a few key differences. Let's get started.

Organizing Certificate Files

To keep things organized, we'll create two new folders:

1. **Backup for old certificates:** This folder will safely store the current SSL certificates you've been using: `~/Documents/GitHub/ deploy-secure-ds-apps-book/config/backup_certificates/ old_certificates`

2. **Storage for multi-domain certificates:** This folder will store the new certificates: ~/Documents/GitHub/deploy-secure-ds-apps-book/config/backup_certificates/new_certificates

Creating the `csr.conf` File

Within the new_certificates folder, create a new file called csr.conf. This configuration file will define the parameters for the CSR, including the primary domain and subdomains. Paste the content from Code-block 6-1 into the file, and modify the fields under [dn] and [alt_names] to match your use case.

Code-block 6-1. csr.conf file

```
[req]
default_bits = 2048
prompt = no
default_md = sha256
distinguished_name = dn
req_extensions = req_ext

[dn]
C = country                    # Country Code (e.g., US)
ST = city                      # State or Province Name
L = locality                   # Locality Name (City)
O = Organization Name          # Organization Name
OU = Department                # Organizational Unit
CN = example.com               # Common Name (Main Domain)

[req_ext]
subjectAltName = @alt_names

[alt_names]
DNS.1 = example.com
DNS.2 = streamlit.example.com
DNS.3 = flask.example.com
DNS.4 = jenkins.lb-webserver.pro
```

Once saved, your `csr.conf` file should look like Figure 6-6.

Figure 6-6. *csr.conf file*

Generating the CSR

Now, let's create the CSR using the OpenSSL command-line tool. Navigate to the new_certificates folder in the terminal and run the command from Code-block 6-2.

Code-block 6-2. Create the CSR

```
cd ~/Documents/GitHub/deploy-secure-ds-apps-book/config/backup_
certificates/new_certificates
openssl req -new -newkey rsa:2048 -nodes -keyout private.key -out example_
com.csr -config csr.conf
```

This command will generate two files within the new_certificates folder:

1. example_com.csr: The certificate signing request file, which you'll use to request the SSL certificate

2. private.key: The private key associated with the certificate

Verifying the CSR

After generating the CSR, it's important to verify that all the specified subdomains were included. Run the following command from Code-block 6-3 in the terminal.

Code-block 6-3. Verify the CSR created

```
openssl req -text -noout -verify -in example_com.csr
```

The output should list all the subdomains you specified in the `csr.conf` file. Compare the output to ensure nothing was missed. See Figure 6-7 for an example of the verification step.

Next Steps

With your CSR file ready, you can now proceed to upload it to the SSL provider (e.g., Namecheap) as part of the certificate issuance process. Once issued, you'll be able to install the multi-domain SSL certificate on your server and secure your applications.

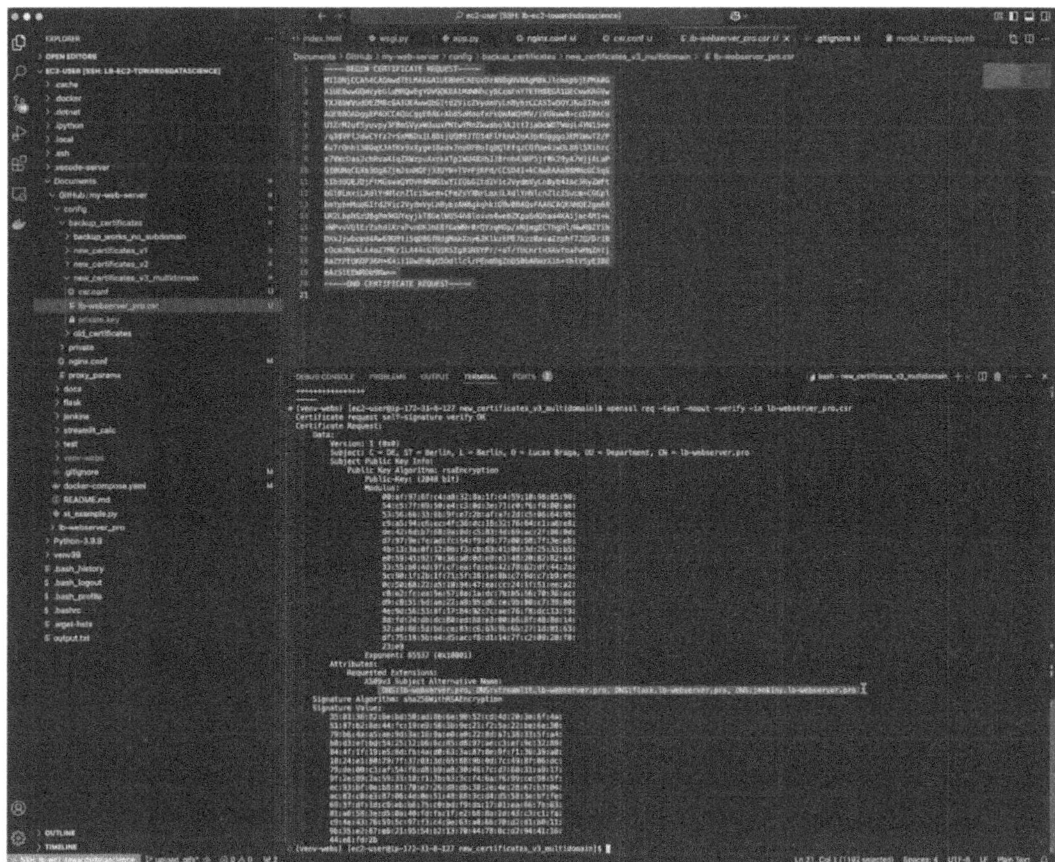

Figure 6-7. *Verify subdomains in CSR*

Activate the SSL Certificate: From ACTIVATE to PENDING

With the CSR created, the next step is to activate your SSL certificate on Namecheap.
Follow the instructions below to proceed.

1. **Access the SSL activation page**

   - Log in to your Namecheap account and navigate to the **SSL
     Certificates** section on the left panel.

   - Locate the certificate you purchased and click **Activate**.

2. **Paste the CSR**

   - On the activation page, you'll see a field to paste your certificate
     signing request (CSR). Open the CSR file you just created, copy its
     contents, and paste it into the provided field on Namecheap.

   - The system will autofill the remaining fields based on the
     information provided in the `csr.conf` file (e.g., country, state,
     and domain names).

     See Figure 6-8.

3. **Verify SANs**

   - If you're following the same setup as this book, you'll have

     - **One primary domain:** `lb-webserver.pro`

     - **Three subdomains:** `streamlit.lb-webserver.pro`, `flask.
       lb-webserver.pro`, and `jenkins.lb-webserver.pro`

   - This makes a total of **four SANs (Subject Alternative Names)**.
     If you've already purchased the additional SAN required for this
     setup, you'll see all the subdomains autofilled as expected.

   - If you haven't purchased the additional SAN yet, you can still do
     so at this step by clicking the **Buy More** option.

4. **Proceed to the next step**

   - Once all the fields are correctly populated and verified,
     click **Next**.

Domain Control Validation

On the next screen, you'll be prompted to confirm the domain control validation (DCV) method. Select **Add CNAME Record** as the validation method and click **Next**.

Confirm Email and Submit

- You'll then be asked to confirm the email address where the SSL files will be sent. Verify that the email address is correct and click **Next**.

- Finally, review all the details:

    - Ensure that the primary domain and all subdomains are listed correctly.

    - Confirm the information and click **Submit** to complete the activation process.

At this point, the status of your SSL certificate will change to **Pending**. This means the SSL provider is waiting for you to complete the validation step, which involves adding the required CNAME records to your DNS settings.

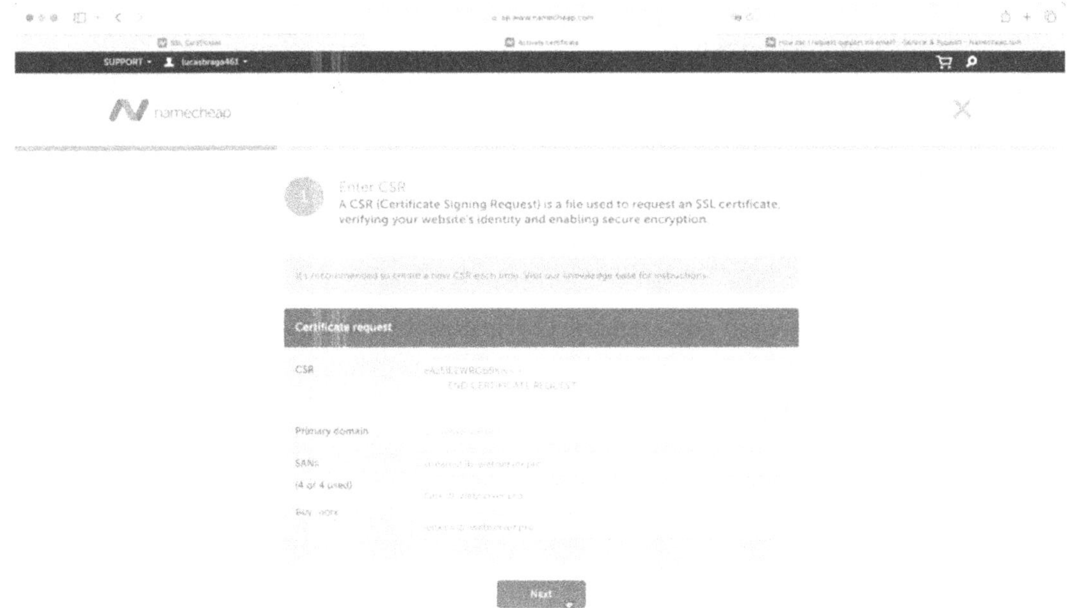

Figure 6-8. *Pasting the CSR into Namecheap*

Activate the SSL Certificate: From PENDING to ACTIVE

Now that the certificate activation process has successfully initiated, the next step is to move its status from **PENDING** to **ACTIVE**. This involves verifying domain ownership by updating the DNS records, a process similar to what we did in Chapter 4, section "Set Up the SSL Certificates." Follow these steps carefully.

1. **Access the certificate details**

   - Navigate to the **SSL Certificates** section on the left panel of the Namecheap dashboard.

   - Locate your new certificate and click **DETAILS**.

   - Scroll down to the bottom of the page and click **See CSR Code**.

   - In the drop-down menu under **EDIT METHODS**, select **Get Record**. A pop-up window will appear, which we'll refer to as the **Get Record tab**.

2. **Update the CNAME Record**

   In the **Get Record tab**, you'll see two pieces of information:

   1. **Host key**

   2. **Target key**

   - Copy the **host key** (omit the suffix, as all keys are identical except for the suffix); see Figure 6-9.

   - Open a new tab and navigate to **Domain List ➤ MANAGE ➤ Advanced DNS** on Namecheap. This tab will be referred to as the **Advanced DNS tab**.

   - Locate the existing CNAME Record and replace the current **host key** with the new one you copied from the **Get Record tab**.

   Next:

   - Copy the **target key** from the **Get Record tab** and paste it into the **target field** of the CNAME Record on the **Advanced DNS tab**, replacing the old value.

   - Set the **TTL field** to the minimum value (one minute) and click the green check mark ✓ to save your changes.

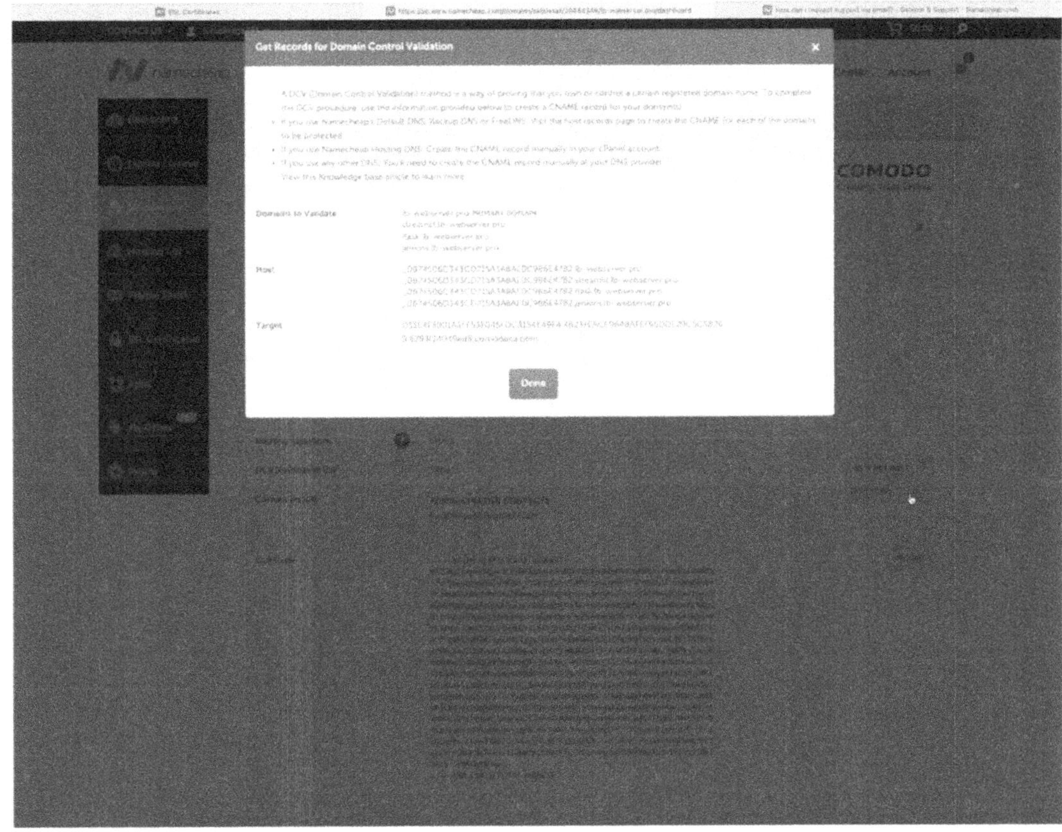

Figure 6-9. *Pasting CSR into Namecheap*

3. **Add A Records for subdomains**

   In Chapter 4, you added an A Record with the following
   configuration:

   - **Host:** @

   - **Value:** <Public IP of the Load Balancer>

That record remains unchanged. Now, for the subdomains, you'll need to add three
additional A Records:

1. **Host:** streamlit

   Value: <Public IP of the Load Balancer>

2. **Host:** flask

   Value: <Public IP of the Load Balancer>

3. **Host:** `jenkins`

   Value: `<Public IP of the Load Balancer>`

Ensure all A Records use the same public IP address of your load balancer. See Figure 6-10 for a visual representation of the DNS configuration.

At this point, your DNS changes will propagate, and the SSL provider will validate the domain ownership. This process may take a few minutes to complete, depending on the TTL setting and DNS propagation times. Once verified, the status of your certificate in the Namecheap dashboard will change to **ACTIVE**.

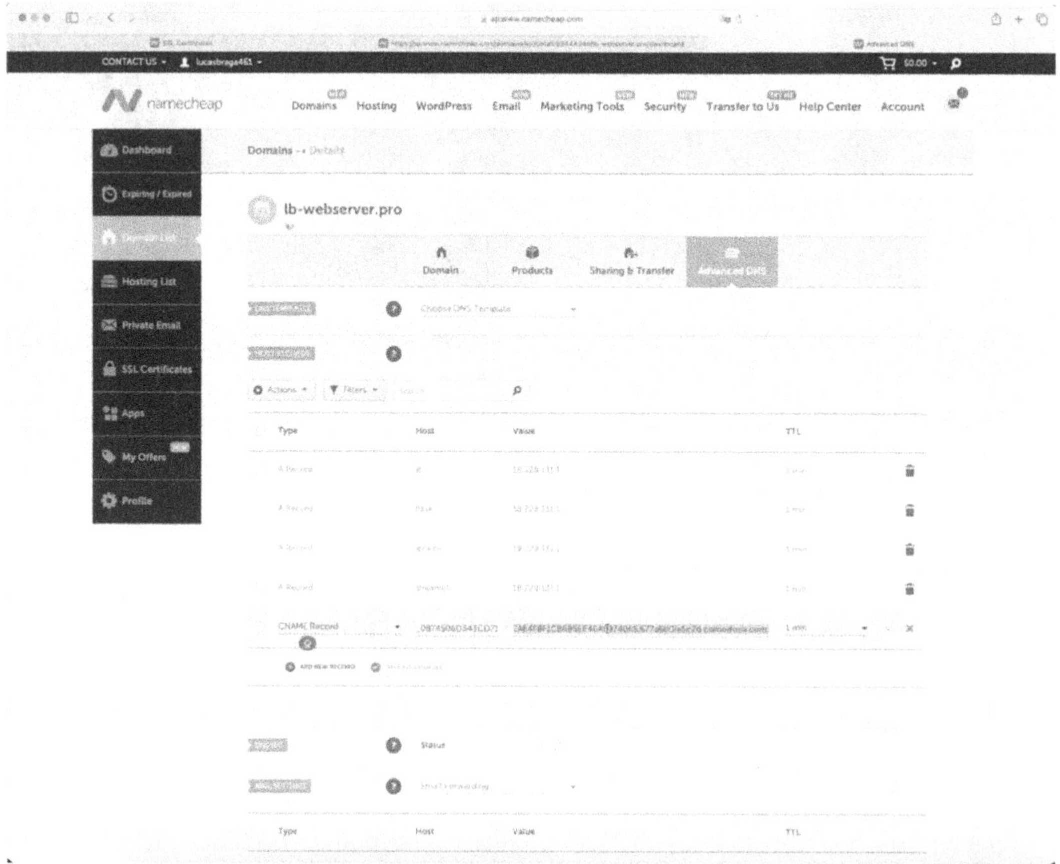

Figure 6-10. *Adjusting the advanced DNS settings*

Prepare the Certificates

At this stage, you should have received an email containing the new SSL certificate files. As with the process described in Chapter 4, you can also download these files directly from the **SSL Certificates** section in the Namecheap console (left panel). Once downloaded, unzip the certificate package, which will include the following files:

- **example_com.ca-bundle** (e.g., `lb-webserver_pro.ca-bundle`)

- **example_com.crt**

- **example_com.p7b**

Next, move these files to the `new_certificates` folder you created earlier in this chapter:

- `~/Documents/GitHub/deploy-secure-ds-apps-book/config/
  backup_certificates/new_certificates`

Combine the Certificate Files

To prepare the certificates for use, follow these steps (similar to the process in Chapter 4), as shown in Figure 6-11.

1. Inside the `new_certificates` folder, create a new file named `example_com_combined.crt` (replace `example_com` with your domain, e.g., `lb-webserver_pro_combined.crt`).

2. Open the **example_com.crt** file, copy its content, and paste it into the newly created `example_com_combined.crt` file.

3. Next, open the **example_com.ca-bundle** file, copy its content, and paste it on a new line at the end of the `example_com_combined.crt` file.

Organize the Files

Once the combined certificate is prepared, ensure your `new_certificates` folder contains the following structure:

- `~/Documents/GitHub/deploy-secure-ds-apps-book/config/backup_
  certificates/new_certificates/example_com_combined.crt`

- `~/Documents/GitHub/deploy-secure-ds-apps-book/config/
  backup_certificates/new_certificates/private/private.key`

This structure will ensure your certificates and private keys are organized and ready for use in the next steps.

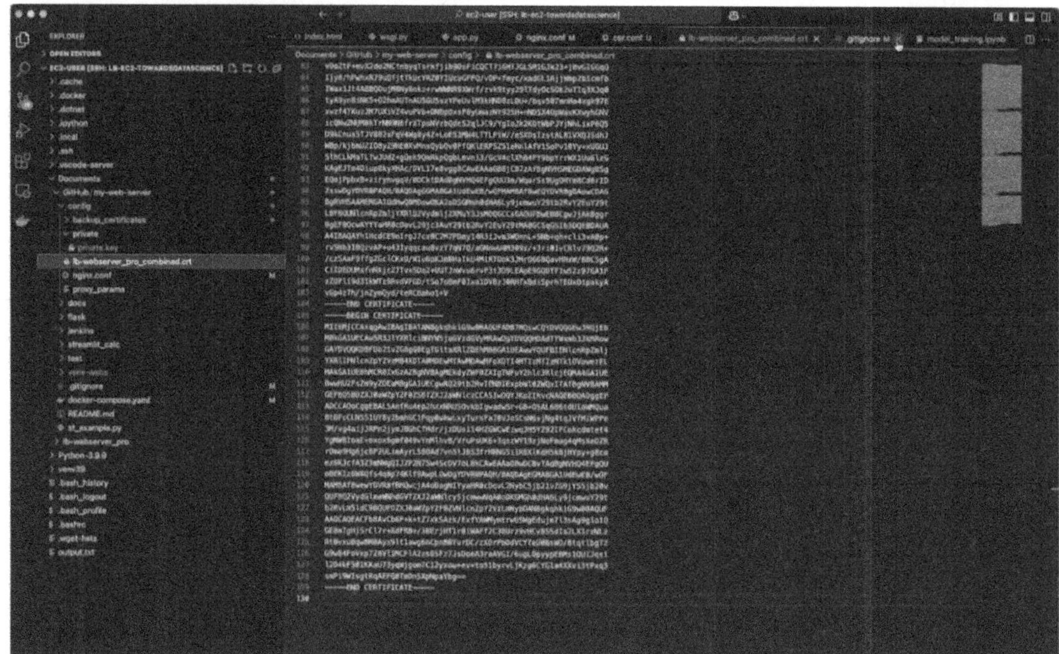

Figure 6-11. *Prepare the certificates*

Deploy the New Changes

Deployment

At this point, you've completed all the necessary setup steps:

- **Nginx and Docker Compose** configurations in the section "Set Up Nginx and Docker-Compose"

- **Adjustments to AWS security groups and load balancer target groups** in the section "Set Up Security Groups and Target Groups on AWS"

- **Acquisition and preparation of new SSL certificates** in the section "Multi-subdomain with SSL Certificate"

Now, let's put all these changes into action by deploying the updated configuration.

Redeploy with Docker Compose

1. Open **VS Code** and navigate to your project repository folder.

2. If Docker is currently running, stop all containers to ensure the updated configuration is applied cleanly.

3. Use the docker-compose command to redeploy your application with the new changes. See Code-block 6-4 and Figure 6-12.

Code-block 6-4. Deploy new changes

```
cd ~/Documents/GitHub/deploy-secure-ds-apps-book
docker-compose down
docker-compose up -d –build
```

Verify the Deployment

After running the above commands, your applications should now be live and accessible through their respective subdomains. Visit each of the following URLs to confirm that the deployment was successful:

- Flask app: flask.example.com (e.g., flask.lb-webserver.pro), shown in Figure 6-13

- Streamlit dashboard: streamlit.example.com (e.g., streamlit.lb-webserver.pro)

- Jenkins server: jenkins.example.com (e.g., jenkins.lb-webserver.pro)

Each subdomain should now be secured with HTTPS and display the expected application.

Figure 6-12. *Running docker-compose on the VS Code terminal*

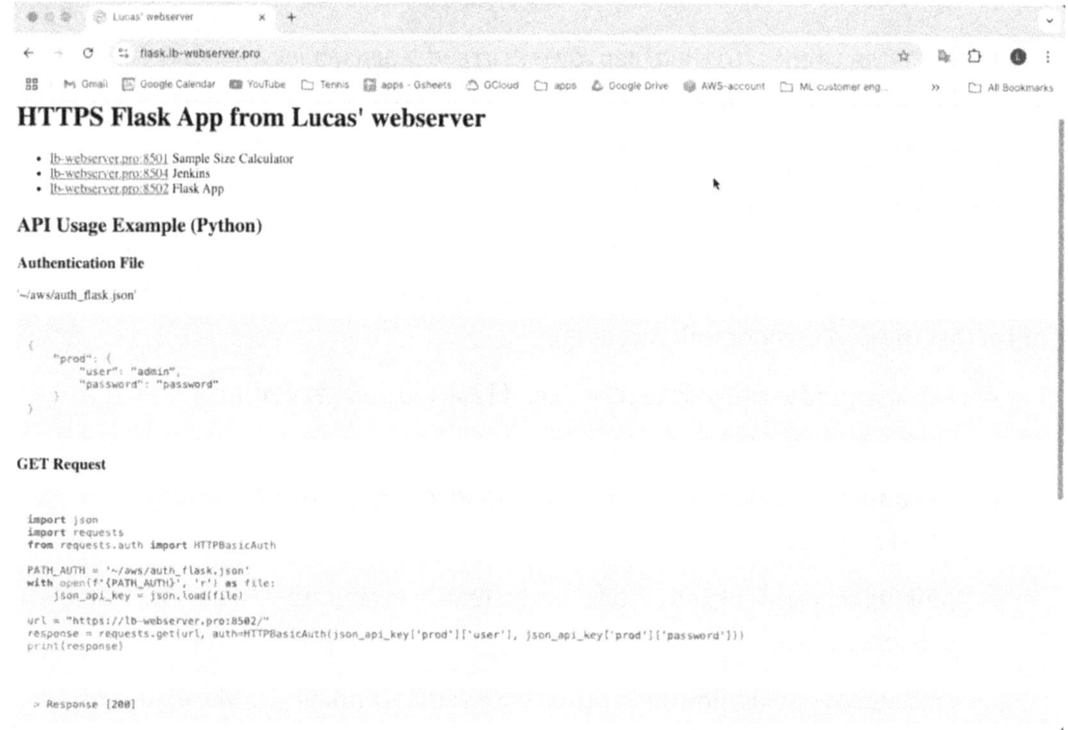

Figure 6-13. *Deployment successful*

Fix Jenkins

After deploying the new configuration, you might notice that jenkins.lb-webserver.
pro is running significantly slower compared to flask.lb-webserver.pro and
streamlit.lb-webserver.pro. This slowdown occurs because Jenkins is still configured

to use its old URL (`https://lb-webserver.pro:8504/`), which no longer matches the new reverse proxy URL (`https://jenkins.lb-webserver.pro`). To resolve this, follow the steps below to update Jenkins' URL configuration:

Steps to Fix Jenkins URL

1. **Log in to Jenkins**

   Access Jenkins through its new URL: `https://jenkins.lb-webserver.pro`.

2. **Navigate to system configuration**

   - Go to **Manage Jenkins ➤ System**.

   - Scroll down to the section labeled **Jenkins Location**.

3. **Update Jenkins URL**

   - Under **Jenkins URL**, replace the old URL (`https://lb-webserver.pro:8504/`) with the new one: `https://jenkins.lb-webserver.pro`; see Figure 6-14.

4. **Save the changes**

   - Click the **Save** button to apply the updated configuration.

Verify the Fix

Once the URL is updated, you should notice an immediate improvement in Jenkins' responsiveness when navigating through the interface. The mismatch between the reverse proxy and Jenkins URL is now resolved, allowing smoother operation.

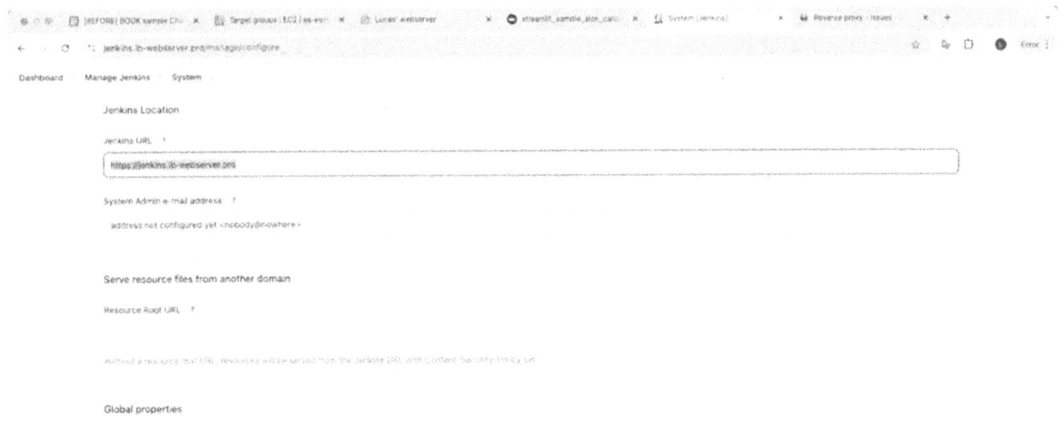

Figure 6-14. *Adjusting Jenkins's new URL*

Summary

In this chapter, you structured and secured your deployment using subdomains. Specifically, you

- Configured subdomains ("flask.", "streamlit.", "jenkins.") using Nginx's reverse proxy ("nginx.conf") and adjusted "docker-compose.yaml" to expose ports 80 and 443 instead of individual app ports

  - Nginx Reverse Proxy documentation: `https://docs.nginx.com/nginx/admin-guide/web-server/reverse-proxy/`

- Updated AWS security groups to allow HTTP (80) and HTTPS (443) traffic and removed unneeded ports (8501, 8502, 8504)

- Reconfigured AWS target groups and load balancer listeners to forward requests from ports 80 and 443 to the correct container services

- Purchased and activated a Multi-domain SSL Certificate from Namecheap to support the main domain and three subdomains

- Generated a certificate signing request (CSR) for multiple subdomains using OpenSSL and a custom "csr.conf" file

  - OpenSSL documentation: `https://docs.openssl.org/1.1.1/man1/req/`

- Configured DNS A Records and CNAME validation via Namecheap's Advanced DNS panel to complete domain control validation

  - Namecheap DNS Management documentation: `https://www.namecheap.com/support/knowledgebase/article.aspx/319/2237/how-can-i-set-up-an-a-address-record-for-my-domain/`

- Prepared certificate files by combining ".crt" and ".ca-bundle" into a single file and organized them for the use by Nginx

- Deployed the updated stack using Docker Compose with HTTPS-enabled subdomains

- Fixed Jenkins by updating its internal URL to match the new subdomain routing ("jenkins.lb-webserver.pro")

In the next chapter, you'll replicate your AWS deployment architecture on Google Cloud Platform (GCP). So all learnings so far from Chapters 1 to 6 that were shown in AWS will be shown in GCP. That's from creating IAM roles and permissions, to launching a virtual machine with static IP, and connecting to it via SSH, and deploying a simple app using Docker to then configuring and deploying the infrastructure with SSL certificates and subdomain routing.

## CHAPTER 7

How to Set Up This Infrastructure on Google Cloud Platform (GCP)

In Chapters 1 through 6, you built a comprehensive, production-ready infrastructure on AWS, covering foundational setup, security best practices, domain and subdomain management, SSL certificates, deployment automation, and container orchestration using Docker and Nginx. Specifically, you deployed Streamlit dashboards, Flask APIs with authentication, and Jenkins automation tools and trained a machine learning model for API integration.

In Chapters 7 and 8, you'll replicate and expand this infrastructure on Google Cloud Platform (GCP). This chapter will guide you through the basics: creating a GCP account and project, configuring IAM roles and permissions, launching a virtual machine with a static IP, and progressively deploying your applications, initially a basic Streamlit app, followed by full integration with SSL certificates, and finally setting up subdomains for organized access.

Chapter 8 will build upon this chapter, introducing more advanced GCP features like load balancing across global regions, autoscaling, creating instance templates and managed instance groups, and deploying strategies to streamline faster updates and scalability.

By the end of these two chapters, you'll have successfully migrated your robust, secure, and scalable infrastructure from AWS to GCP, leveraging the unique capabilities and efficiencies offered by Google's cloud environment.

© Lucas H. Benevides e Braga 2025
L.H.B.e. Braga, *Deploying Secure Data Science Applications in the Cloud*,
https://doi.org/10.1007/979-8-8688-1715-1_7

Creating a GCP Account, Project, and IAM User Role

Follow these detailed steps to set up your GCP Platform environment:

1. **Log in to Google Cloud Console**

   - Visit console.cloud.google.com and sign in with your personal or corporate Google account; see Figure 7-1.

2. **Create a new project**

   - On the top-left corner, click **Select a project**.

   - In the pop-up window that appears, click **New Project** located at the top right.

   - Enter a descriptive name for your project.

   - If you're using a corporate account, select the appropriate organization from the drop-down menu.

   - Click **Create** to finalize the project setup.

Note   Google often provides new users with a $300 credit for initial experimentation. Check your eligibility when setting up your account. Figure 7-2 shows what appears on your screen when you're eligible for the 300 USD credits from GCP.

---

3. **Selecting your project**

   - After creating the project, return to the top-left corner and click **Select a project** again.

   - From the list, choose the project you just created.

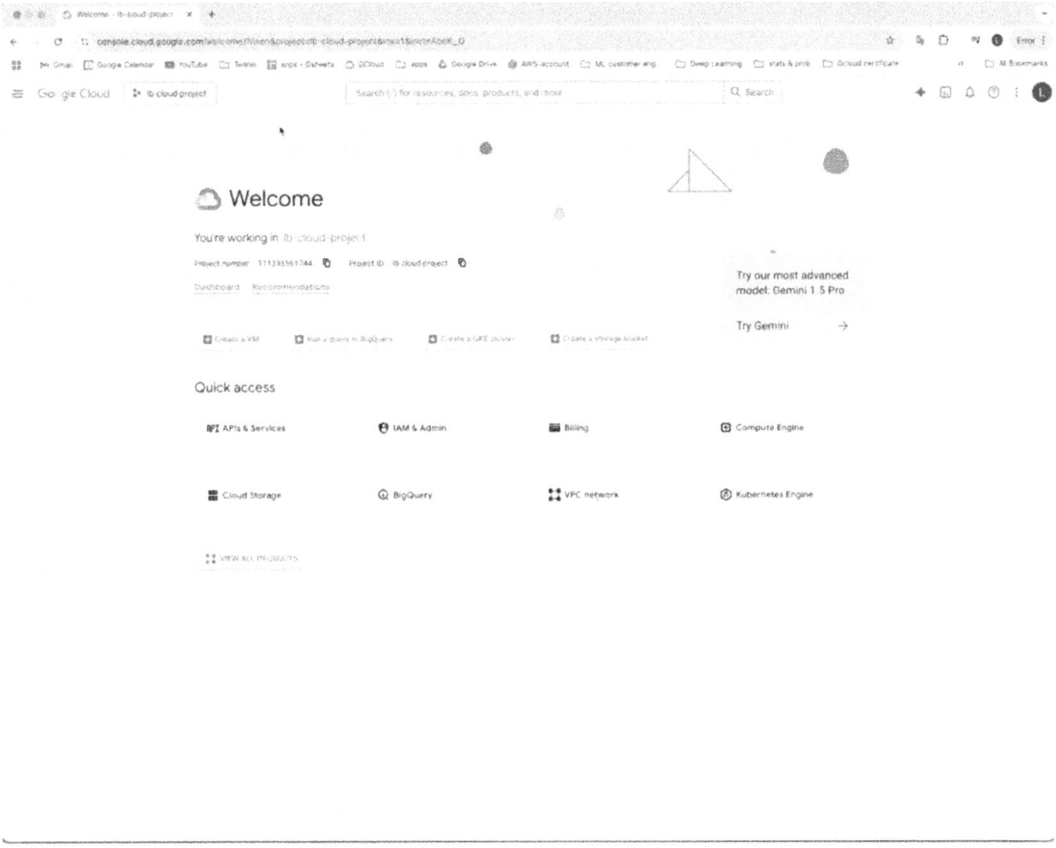

Figure 7-1. *GCP console*

4. **Configuring IAM user roles**

- Click the **Navigation Menu** (three horizontal lines icon) on the top left.

- Navigate to **IAM & Admin.**

- In the **IAM & Admin** page, click the **GRANT ACCESS** button.

- A panel will appear on the right side. In the **New principals** field, enter the email address of the user you wish to grant access to.

- Click **Select a role** and choose the following role to ensure the user has the necessary permissions:

  - **Compute Engine ➤ Compute Admin**

See Figure 7-3.

Your Google Cloud project is now set up with the appropriate IAM user role configured.

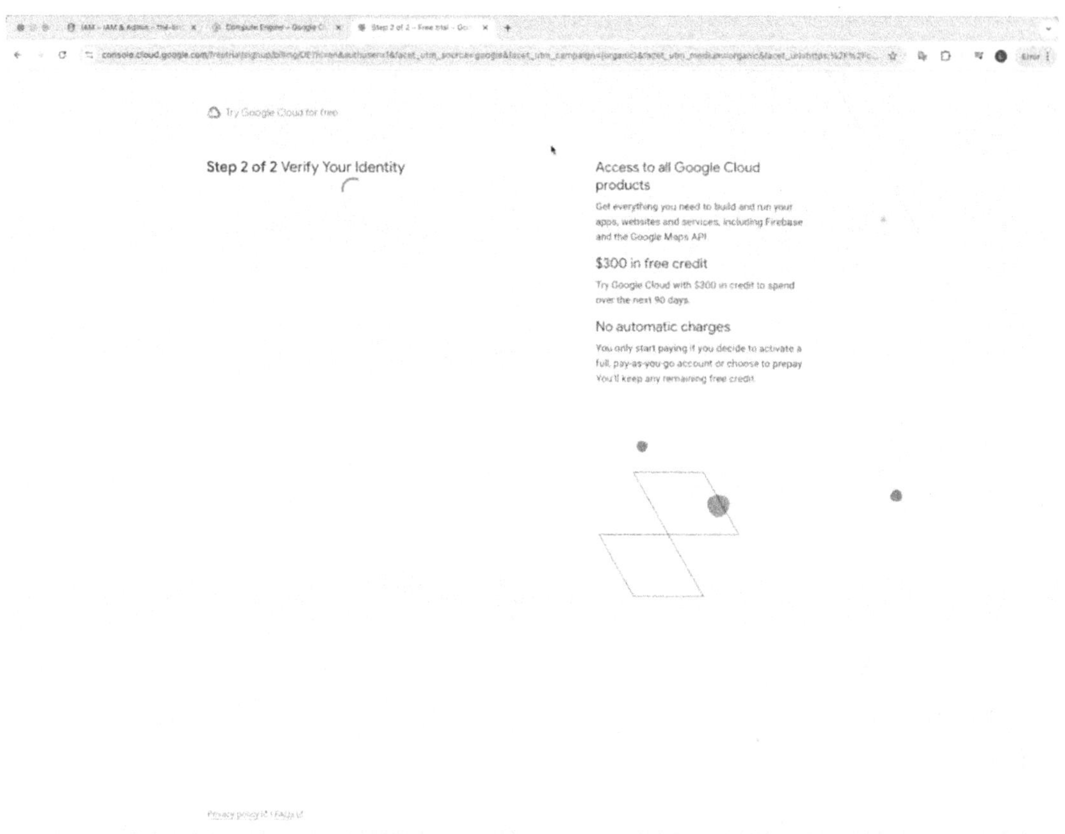

Figure 7-2. *Eligible for 300 USD in credit*

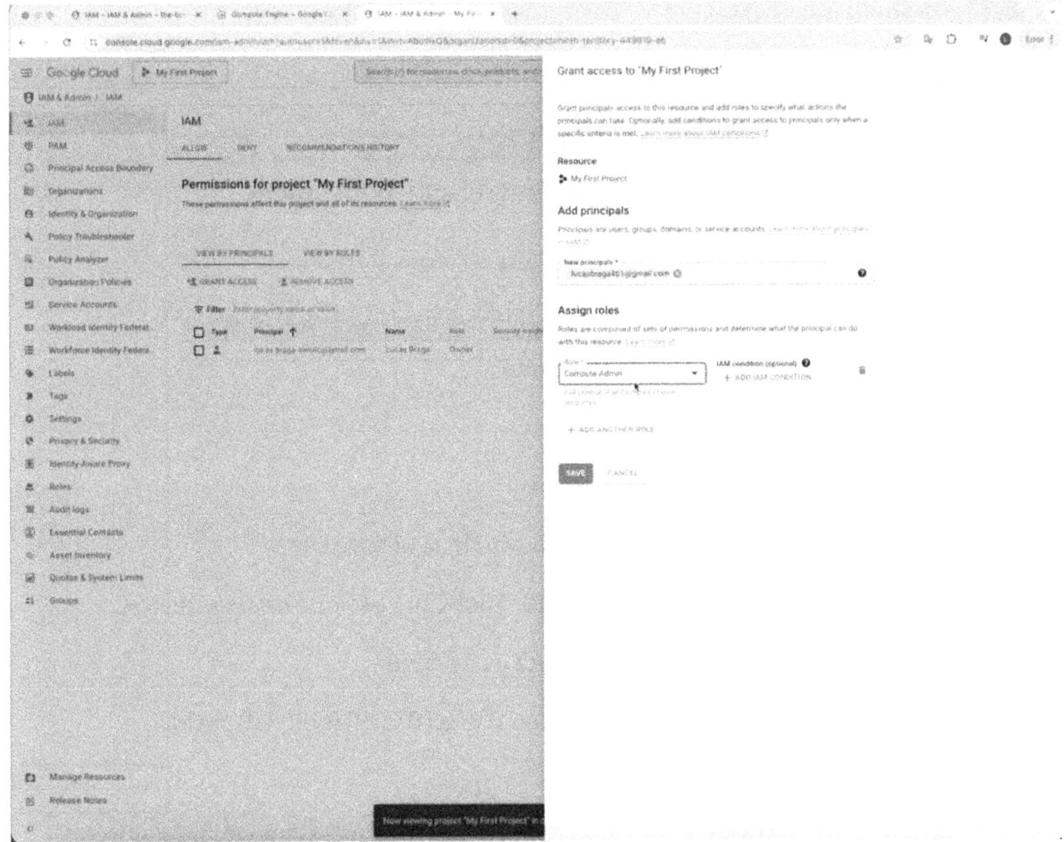

Figure 7-3. *Grant user Compute Admin role*

Launching a Virtual Machine (VM) with a Startup Script

This section will guide you through creating a virtual machine (VM) on Google Cloud Platform using an automated startup script. This approach streamlines the setup process, ensuring reproducibility and minimizing manual errors and dependency conflicts.

Follow these detailed steps:

1. **Navigate to Compute Engine**

    - Click the **Navigation Menu** (top left).

    - Hover over **Compute Engine** and select **VM Instances.**

2. **Enable Compute Engine API**

    - If prompted, click **Enable** to activate the Compute Engine API.

3. **Create and configure a VM instance**

    - Click **Create Instance.**

4. **Set up the machine configuration**

    - **Name:** Enter a descriptive name such as `webserver`.

    - **Region:** Choose `europe-west10 (Berlin)` or your nearest region.

    - **Zone:** Leave the default or select as needed.

    - **Machine type:** Choose `e2-standard-8` (8 vCPUs, 32 GB RAM).

5. **Configure OS and storage**

    - Click the **CHANGE** button under **Operating system and storage**.

    - Set **Size (GB)** to 100.

    - Click **SELECT** to confirm.

6. **Configure networking**

    - In the Networking section, set the **Network tags** to `vm-network-tag`.

7. **Add the startup script**

    - Click **Advanced options** at the bottom.

    - Under the **Automation** tab, paste the following script into the **Startup script** field, from Code-block 7-1. The script is available at the book's repository in the right format to be applied.

Code-block 7-1. Startup script for VM instance in Google Cloud

```bash
#!/bin/bash

# Update and upgrade the system
sudo apt update -y
sudo apt upgrade -y

# Install Python and essential packages
sudo apt install -y python3 python3-pip python3-venv

# Install Docker
sudo apt install -y apt-transport-https ca-certificates
curl gnupg
curl -fsSL https://download.docker.com/linux/debian/gpg | sudo
gpg --dearmor -o /usr/share/keyrings/docker-archive-keyring.gpg
echo "deb [arch=$(dpkg --print-architecture) signed-by=/usr/
share/keyrings/docker-archive-keyring.gpg] https://download.
docker.com/linux/debian $(lsb_release -cs) stable" | sudo tee /
etc/apt/sources.list.d/docker.list > /dev/null
sudo apt update -y
sudo apt install -y docker-ce docker-ce-cli containerd.io

# Start and enable Docker service
sudo systemctl start docker
sudo systemctl enable docker

# Add the current user to the Docker group
sudo usermod -aG docker $USER

# Install Docker Compose
sudo curl -L "https://github.com/docker/compose/releases/
latest/download/docker-compose-$(uname -s)-$(uname -m)" -o /
usr/local/bin/docker-compose
sudo chmod +x /usr/local/bin/docker-compose
```

```
# Install Git
sudo apt install -y git

# Verify installations
echo "Installed versions:"
python3 --version
docker --version
docker-compose --version
git --version
```

8. **Finalize and create the VM**

   - Click **Create** at the bottom of the page to deploy your VM
     instance.

Once you've clicked **Create**, Google Cloud will begin provisioning your new virtual machine. This process might take some minutes, depending on the selected configurations and the region chosen.

Detailed Explanation of the Startup Script

The startup script you added earlier automates several critical setup tasks, which were initially introduced in Chapter 2 through manual step-by-step instructions. Automating these tasks enhances reproducibility and consistency and reduces potential errors due to manual operations. Here's exactly what the script accomplishes:

- System Update and Upgrade

- Python and Essential Packages Installation

- Docker Installation

- Docker Service Setup

- Docker Compose Installation

- Git Installation

- Verification Step

Automating these steps through a startup script significantly streamlines the initial VM setup, making it consistent and error-free. It eliminates manual errors, ensures identical environments across deployments, and addresses potential compatibility or dependency conflicts up front.

Assigning a Static IP and Connecting to the VM via SSH

SSH Connection Using VS Code

Follow these steps carefully to set up a secure SSH connection between your local machine (using VS Code) and your newly created Google Cloud VM:

1. **Generate SSH key pair**

   - On your local machine, open the terminal and execute the following command from Code-block 7-2.

     Code-block 7-2. ssh config file to connect to the GCP virtual machine

     ```
     ssh-keygen -t rsa -b 4096 -C "your_email@example.com"
     ```

   - This command generates two files in your ~/.ssh directory:

     - **Public key:** id_rsa.pub

     - **Private key:** id_rsa

2. **Add SSH public key to Google Cloud VM**

   - Open the public key file (~/.ssh/id_rsa.pub) and copy its content.

   - Navigate to your Google Cloud Console:

     - Go to **Compute Engine ➤ VM Instances**.

     - Click your VM instance, then click **EDIT**.

     - Scroll down to the **Security and Access** section.

     - Under **SSH Keys**, click **ADD ITEM**.

     - Paste the copied public key into the field provided.

     - Click **SAVE** at the bottom of the page.

   See Figure 7-4.

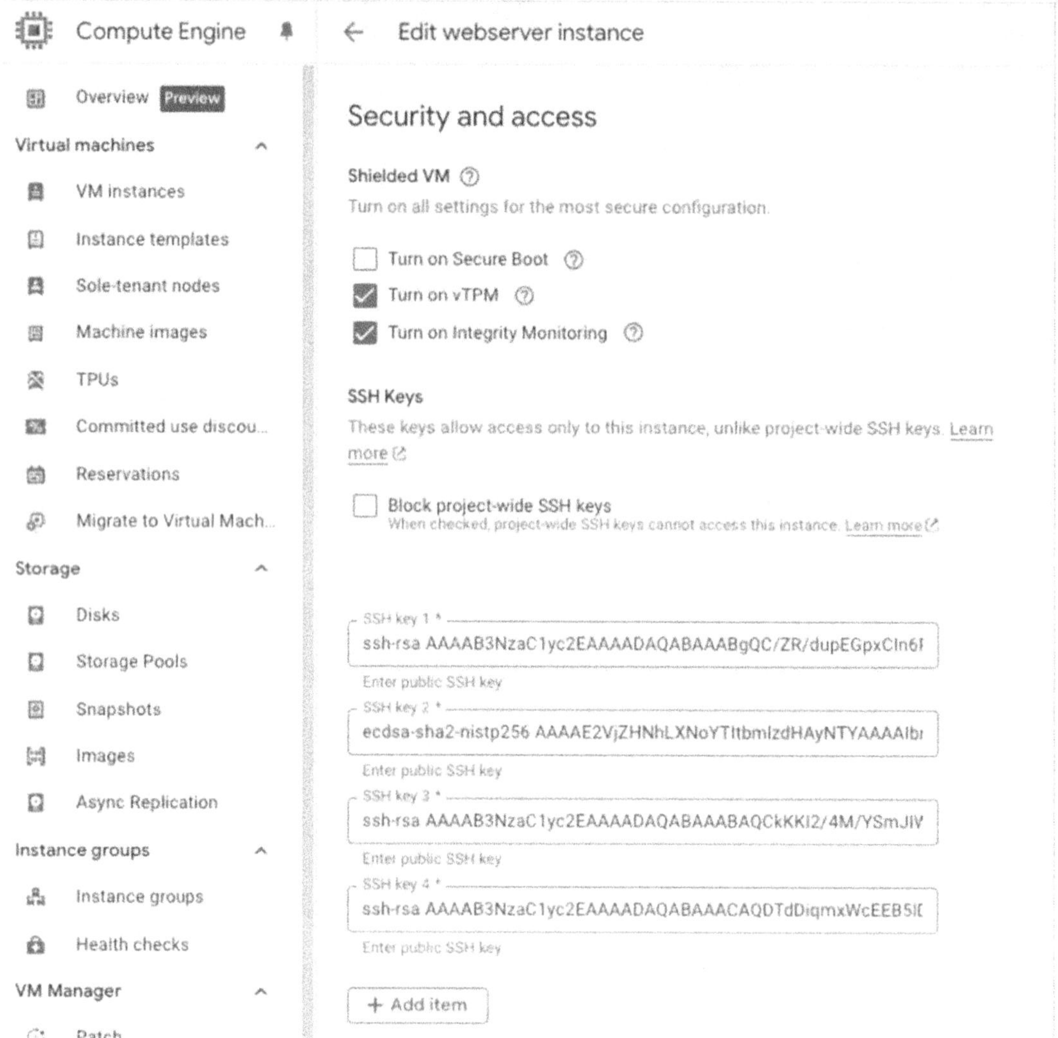

Figure 7-4. *Add SSH keys.*

3. **Reserve an external static IP address**

    - In Google Cloud Console, navigate to

        - **VPC Network ➤ IP Addresses.**

        - Click **RESERVE EXTERNAL STATIC IP ADDRESS.**

- Fill out the form with the following details:

  - **Name:** A descriptive name, for example, `static-ip-webserver`.

  - **Network service tier:** `Standard`.

  - **Type:** `Regional` (select the same region as your VM instance, e.g., `europe-west10`).

  - **Attached to:** Select your existing VM instance, for example, `webserver-vm`.

- Keep all other options at default values and click **RESERVE**.

4. **Configure local SSH setup for VS Code**

   - On your local machine, edit or create the file `~/.ssh/config` with the following configuration (replace `34.1.169.121` with your actual reserved static IP address and `your-username` with your Google Cloud username); see Code-block 7-3 and Figure 7-5.

     Code-block 7-3. ssh config file to connect to the GCP virtual machine

```
Host vm-book-project
    HostName 34.1.169.121
    IdentityFile ~/.ssh/id_rsa
    User lucas.braga.aiesec
```

5. **Establish SSH connection**

   Ensure that the `HostName` field on Code-block 7-3 contains the static IP address you reserved earlier. Your configuration is ready; use the instructions from the section "SSH Connection to the Instance Using VS Code" in Chapter 2. Because we haven't restricted access to port 22 during VM creation, your SSH connection should work successfully.

Note on Firewall Settings   Google Cloud automatically provides a default firewall rule called `default-allow-ssh`, allowing SSH connections on port 22. This default rule applies to all VM instances (unless modified by custom corporate configurations). Be aware of firewall rule priorities (numbered, with lower numbers taking precedence). If another firewall rule with a lower priority (e.g., 1000 or 1) explicitly blocks port 22, SSH connections will be denied despite the default rule.

Ensure no conflicting custom rules exist that may unintentionally restrict your SSH access.

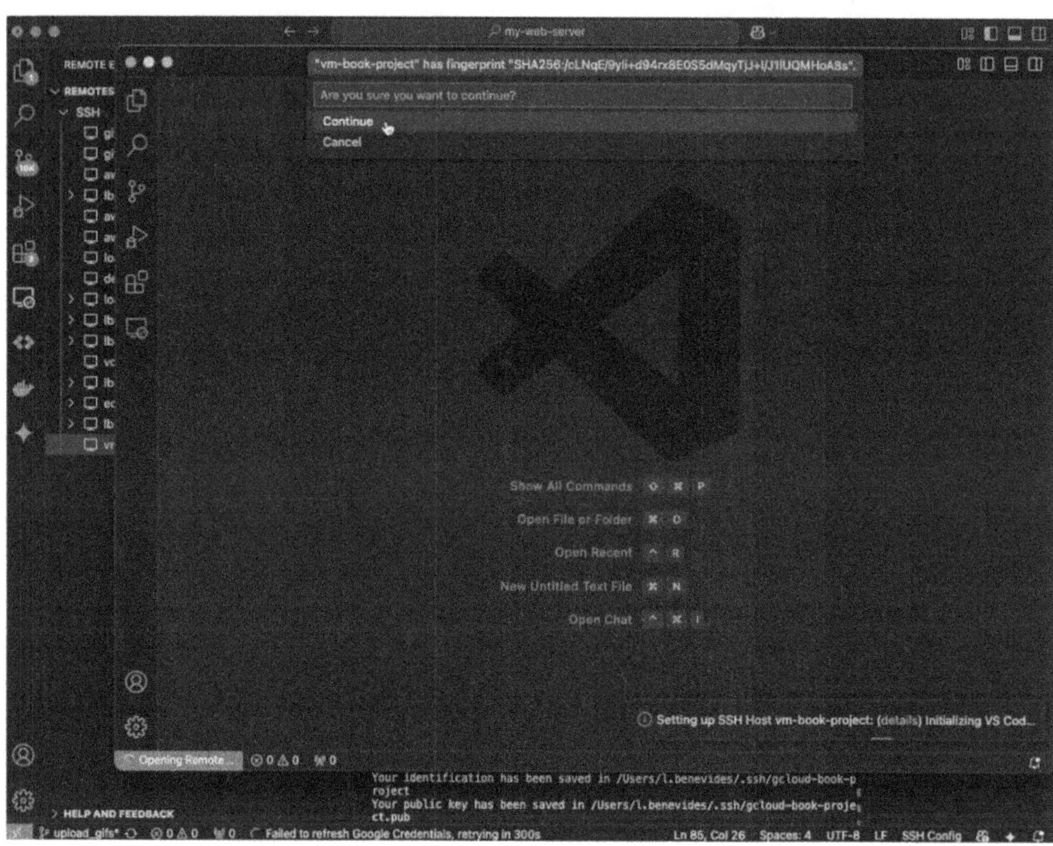

Figure 7-5. *Connecting via SSH using VS Code*

SSH Connection Using GCP Console

An alternative, simpler way to establish an SSH connection to your VM is through the Google Cloud Console. This method provides direct terminal access through your web browser, making it ideal if you prefer simplicity and do not require VS Code integration.

Follow these steps:

1. In the Google Cloud Console, navigate to **Compute Engine ➤ VM Instances**.

2. Locate your VM instance, then click on the **Connect** button next to it, and select **SSH** from the drop-down menu.

3. A new browser window or tab will open, requesting authorization for access. Click **Authorize** to proceed.

You will now have direct access to a terminal window where you can manage your VM using standard Linux bash commands (see Figure 7-6).

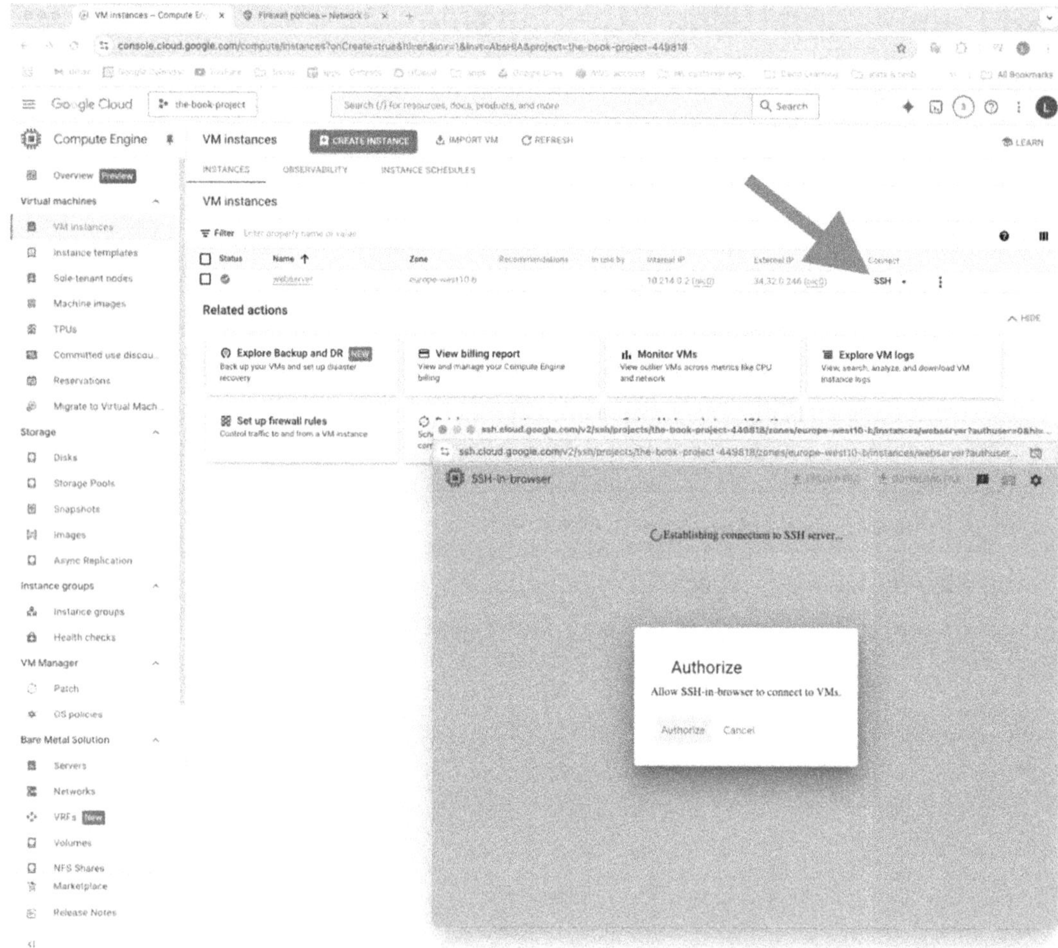

Figure 7-6. *Connecting via SSH using GCP console*

Level 1: Deploy a Simple Streamlit App on Port 8501

Deploy a Simple Streamlit App

Once you've successfully established an SSH connection to your VM, follow these steps to deploy a basic Streamlit application. We'll create a directory, set up a Python virtual environment, install Streamlit, and launch a simple test application.

Step-by-Step Instructions

1. **Create and navigate into the `Documents` folder**; see
   Code-block 7-4.

   Code-block 7-4. Deploy a dummy Streamlit app

```
cd ~
mkdir Documents
cd Documents
```

2. **Create and activate a Python virtual environment** (you may
   name it venv_main or choose another name); see Code-block 7-5.

   Code-block 7-5. Create a virtual environment

```
python3 -m venv venv_main
source venv_main/bin/activate
```

3. **Install Streamlit within the activated virtual environment**; see
   Code-block 7-6.

   Code-block 7-6. Create a virtual environment

```
pip install streamlit
```

4. **Create a new Python file named `st_example.py`** with the
   following Streamlit example code; see Code-block 7-7.

   Code-block 7-7. st_example.py

```python
import streamlit as st

def main():
    st.title("Streamlit App running on Google Cloud via
    port 8501")
    st.write("Test successful!")

if __name__ == "__main__":
    main()
```

If you're using VS Code, you can create and edit this file directly in the editor. Otherwise, if you prefer using the terminal, follow these instructions; see Code-block 7-8.

Code-block 7-8. Create the file st_example.py using the terminal

```
cd ~/Documents
nano st_example.py
# Paste the code provided above (On Mac: CMD+V, on
Windows/Linux: Ctrl+Shift+V or right-click to paste)
# Exit and save: Press Ctrl+X, then Y, and Enter
```

5. **Run your Streamlit application**; see Code-block 7-9.

Code-block 7-9. Run/deploy Streamlit

```
streamlit run st_example.py
```

After running this command, you should see output similar to the following:

- Local URL: http://localhost:8501

- Network URL: http://10.214.0.2:8501

- External URL: http://<YOUR_STATIC_IP>:8501 (in my case http://34.1.169.121:8501)

Understanding URLs and Accessing Your App

- **Local URL:** Accessible directly if you're connected via VS Code SSH or another method that forwards local ports.

- **Network URL:** Accessible only within the VM's internal network.

- **External URL:** Intended for public access; however, since you have not yet configured Google Cloud's firewall rules to allow ingress traffic on port 8501, this URL won't be reachable at this point.

Important Notes

- **If you're using VS Code's SSH:** You should be able to use the **Local URL** (`http://localhost:8501`) immediately within your local browser because VS Code forwards local ports automatically.

- **If you're connecting via Google Cloud Console SSH:** You probably won't be able to access the Local URL from your browser, as the Google Cloud Console SSH opens a separate browser-based terminal session that doesn't forward the VM's local ports to your local machine. In this case, you'll eventually need to use the **External URL** (once firewall rules are configured properly) to access the Streamlit application from your browser.

Proceed to the next section to configure Google Cloud Firewall rules to enable external access to your Streamlit application.

Firewall Rules with Network Tagging (Allow 8501 Ingress)

To access your Streamlit application externally, you must configure Google Cloud Firewall rules to allow inbound traffic on port 8501. Follow these detailed steps:

1. **Navigate to Firewall settings**

   - In the Google Cloud Console, click the **Navigation Menu**.

   - Select **VPC Network** and then click **Firewall**.

   You'll likely notice that no current firewall rule permits traffic on port 8501 (see Figure 7-7). We'll create a new rule for this purpose.

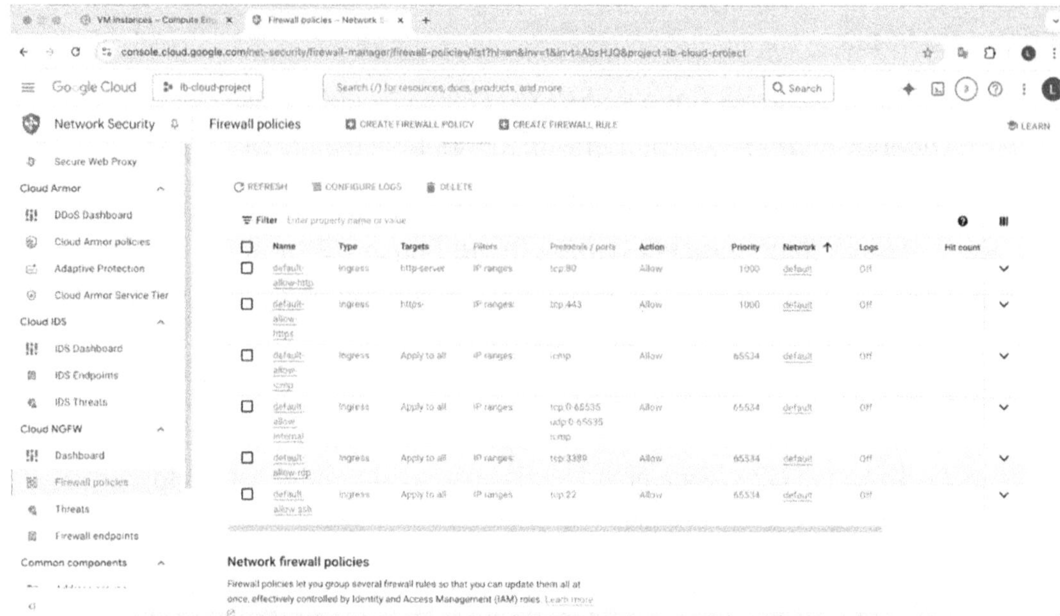

***Figure 7-7.** Firewall rules*

2. **Create a new firewall rule**

   - Click the **CREATE FIREWALL RULE** button.

   - Fill in the fields as follows:

     - **Name:** Provide a descriptive name, such as `allow-rule-webserver`.

     - **Network:** Select the same network used for your VM instance (typically `default`).

     - **Priority:** Set it to 1000. Lower numbers are given higher priority, ensuring this rule is prioritized appropriately.

     - **Action on match:** Choose `Allow` (since this rule should permit traffic).

     - **Target tags:** Enter the network tag (`vm-network-tag`) previously assigned to your VM in the section "Creating a GCP Account, Project, and IAM User Role."

- **Source IPv4 ranges:** Specify IP ranges allowed to access your application. For enhanced security, restrict access to your specific IP by adding your IP address followed by /32. (Find your IP at whatismyipaddress.com, e.g., 87.154.215.35/32.)

- **Protocols and ports:** Select `Specified protocols and ports` and enter `tcp:8501`.

- Once configured, click **CREATE** to apply the rule.

Refer to Figures 7-8 and 7-9 for reference.

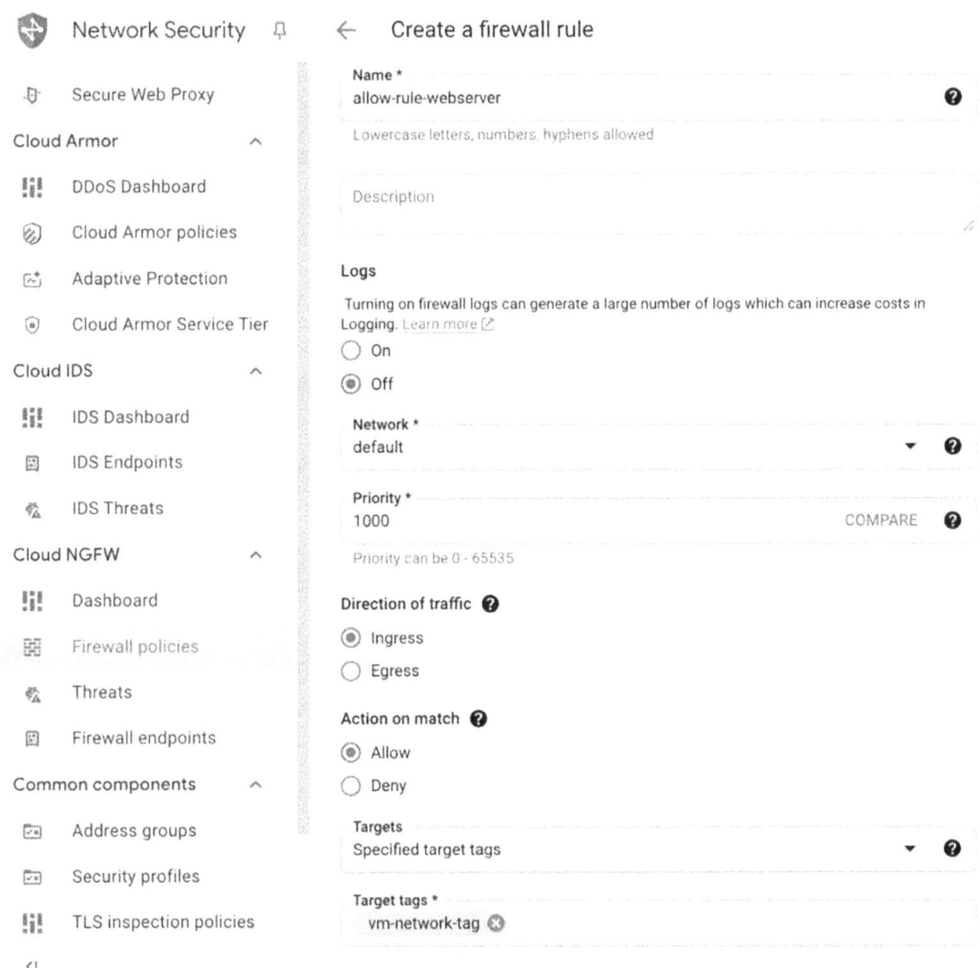

Figure 7-8. *Create Firewall rule—part 1*

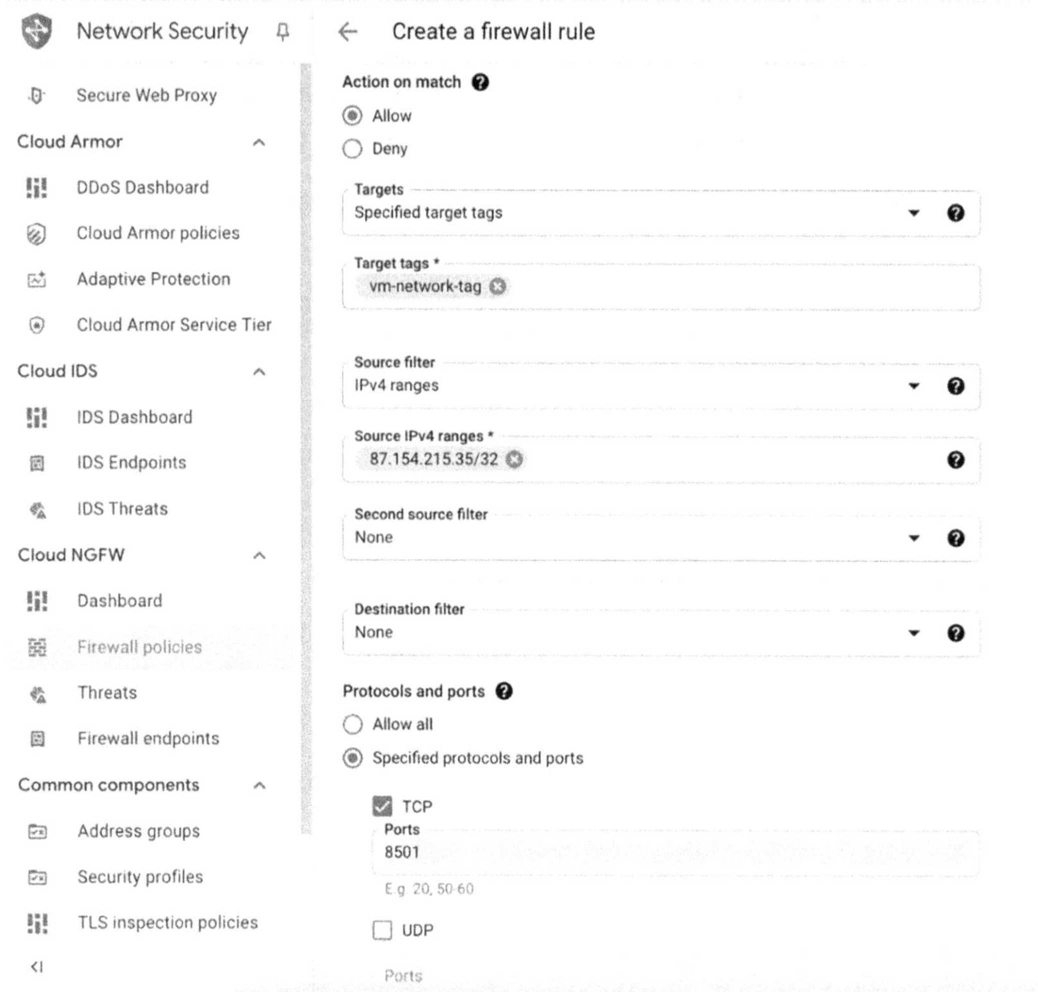

Figure 7-9. *Create Firewall rule—part 2*

3. **Verify external access**

- Now that the firewall rule is established, attempt to access your
  Streamlit app using the previously provided External URL:
  `http://<YOUR_STATIC_IP>:8501`.

4. **Update application text (optional)**

- If you notice the title of your Streamlit application still references
  "ec2", update it to reflect Google Cloud by modifying the line of
  code from `st_example.py`; see Code-block 7-10.

136

Code-block 7-10. Update application text

```
# old:
st.title("Streamlit App running ec2 via port 8501")
# new:
st.title("Streamlit App running on Google Cloud via port 8501")
```

- Save your changes, stop the current Streamlit process by pressing Ctrl+C in your terminal, and restart the application to reflect the new text (see Figure 7-10).

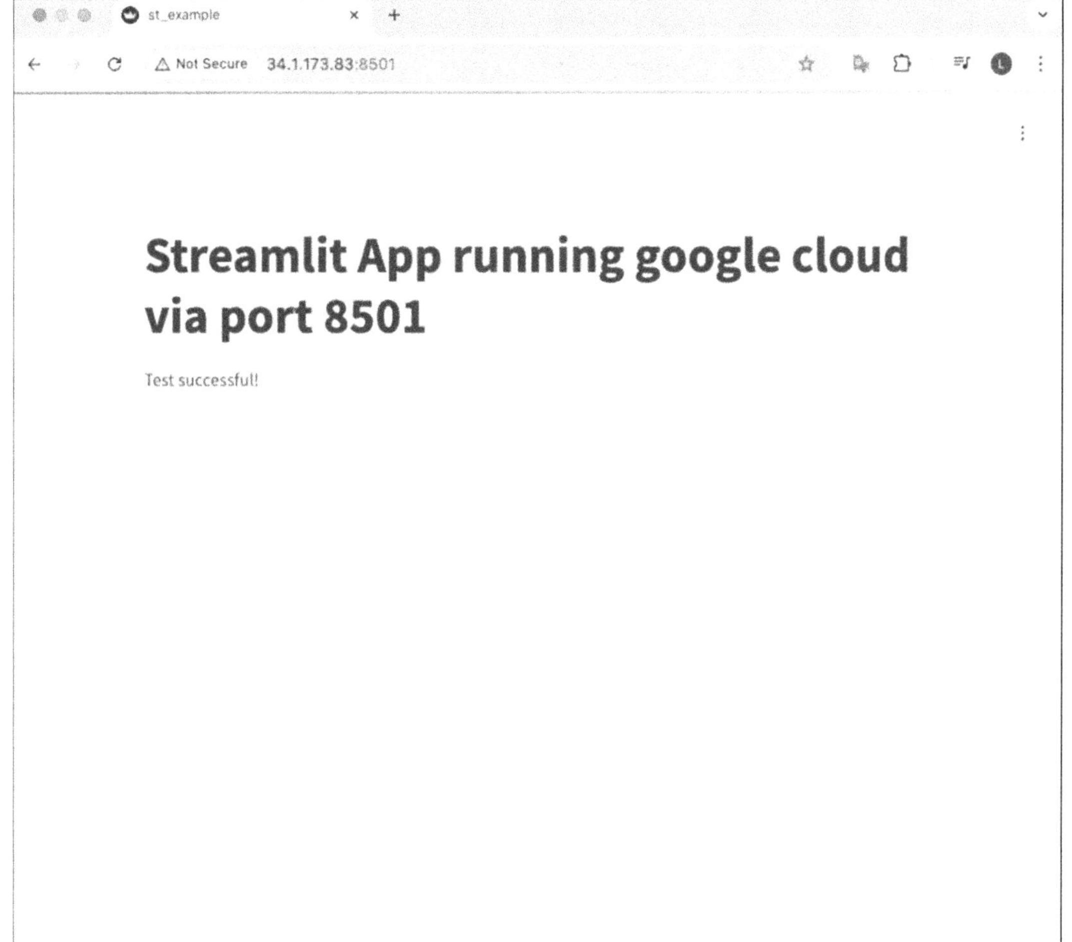

Figure 7-10. *Modified Streamlit title from ec2 to Google Cloud*

Your Streamlit application is now successfully accessible externally, secured by the new firewall rules you've configured.

Level 2: Deploy with SSL Using GitHub Code

Configure GitHub and Clone the Repository

Follow these steps to configure your SSH access to GitHub and clone the required repository. Detailed guidance on this process was provided in the section "Set Up GitHub and Clone the Repository" in Chapter 2; here's a concise summary:

Step-by-Step Instructions

1. **Generate SSH keys (if not previously created)**; see Code-block 7-11.

   Code-block 7-11. Generate SSH keys

   ```
   ssh-keygen -t ed25519 -C "your_email@gmail.com"
   ```

2. **Copy the public key:**

   - On macOS, copy the public key using the command from Code-block 7-12.

   - If pbcopy is unavailable or you're on another OS, manually copy the key from the file located at ~/.ssh/id_ed25519.pub.

     Code-block 7-12. Copy the public key using the terminal

     ```
     pbcopy < ~/.ssh/id_ed25519.pub
     ```

3. **Add the public key to GitHub:**

   - Navigate to your GitHub account settings, select **SSH and GPG keys**, and paste the copied public key.

4. **Verify GitHub SSH Access**; see Code-block 7-13.

   Code-block 7-13. Verify access

   ```
   ssh -T git@github.com
   eval "$(ssh-agent -s)"
   chmod 700 ~/.ssh/
   ```

5. **Clone the GitHub repository**; see Code-block 7-14.

Code-block 7-14. Clone the GitHub repository

```
mkdir -p ~/Documents/GitHub
cd ~/Documents/GitHub
git clone git@github.com:lucasbraga461/deploy-secure-ds-apps-book.git
cd ~/Documents/GitHub/deploy-secure-ds-apps-book
```

6. **Configure Git user identity**; see Code-block 7-15.

Code-block 7-15. Configure Git user

```
git config --global user.email "you@example.com"
git config --global user.name "Your Name"
```

Configure User Permissions

Set up necessary user groups and permissions by applying knowledge from Chapter 5 (see Code-block 7-16); it's recommended to go through the comments to understand the process.

Code-block 7-16. Users and groups setup – learnings from the section "Deploy Jenkins" in Chapter 5

```
# 1. create shared group
cd ~
sudo groupadd jenkins_shared

# 2. add your user to the group, that is the prefix of the
#  email from your public key, in my case lucas.braga.aiesec
sudo usermod -aG jenkins_shared lucas.braga.aiesec

# 3. if docker already exist then:
sudo usermod -aG jenkins_shared docker
# otherwise:
sudo useradd -g jenkins_shared docker
sudo usermod -aG jenkins_shared docker
```

```
# 4. create jenkins user and add to jenkins_shared
sudo groupadd jenkins
sudo useradd -g jenkins jenkins
sudo usermod -aG jenkins_shared jenkins

# 5. create nginx user and add to jenkins_shared
sudo groupadd nginx
sudo useradd -g nginx nginx
sudo usermod -aG jenkins_shared nginx

# 6. change ownership of the jenkins directory
cd ~/Documents/GitHub/deploy-secure-ds-apps-book/jenkins
sudo chown -R $(id -u jenkins):jenkins_shared var/jenkins_home
sudo chmod -R 775 var/jenkins_home

# 7. verify that they all belong to jenkins_shared
groups jenkins
groups docker
groups nginx

# If not add them
sudo usermod -aG jenkins_shared jenkins
sudo usermod -aG jenkins_shared docker
sudo usermod -aG jenkins_shared nginx

# 8. Create a .env file with the UID:GID from jenkins
#  on the same folder as the docker-compose.yaml file
#  these variables will be used by the docker-compose.yaml file
cd ~/Documents/GitHub/deploy-secure-ds-apps-book/
echo "JENKINS_UID=$(id -u jenkins)" > .env
echo "JENKINS_GID=$(id -g jenkins)" >> .env
```

Your docker-compose.yaml should reflect the structure from Code-block 7-17.

Code-block 7-17. docker-compose.yaml file

```
version: "3.7"

services:
 streamlit_calc:
   build:
     context: streamlit_calc
   ports:
     - "8501"
   restart: always

 flask:
   build:
     context: flask
   ports:
     - "8502"
   volumes:
     - ./flask/logs:/app/logs
   restart: always

 jenkins:
   build:
     context: jenkins
   ports:
     - "8080"
   user: "${JENKINS_UID}:${JENKINS_GID}"
   environment:
     - JENKINS_UID=${JENKINS_UID}
     - JENKINS_GID=${JENKINS_GID}
     - AWS_CONFIG_FILE=/app/.aws/config
   volumes:
     - ./streamlit_calc/data:/app/data
     - ./jenkins/.aws:/app/.aws
     - ./jenkins/var/jenkins_home:/var/jenkins_home
   restart: always
```

```
nginx:
  image: nginx:latest
  volumes:
    - ./config/nginx.conf:/etc/nginx/nginx.conf:ro
    - ./config/lb-webserver_pro_combined.crt:/etc/pki/nginx/lb-webserver_
      pro_combined.crt:ro
    - ./config/private/private.key:/etc/pki/nginx/private/private.key:ro
    - ./config/proxy_params:/etc/nginx/proxy_params:ro
  depends_on:
    - streamlit_calc
    - jenkins
    - flask
  ports:
    - "8501:8501"
    - "8504:8504"
    - "8502:8502"
  restart: always
```

Your `nginx.conf` should reflect the structure from Code-block 7-18.

Code-block 7-18. nginx.conf file

```
# For more information on configuration, see:
#   * Official English Documentation: http://nginx.org/en/docs/

user nginx;
worker_processes auto;
error_log /var/log/nginx/error.log;
pid /run/nginx.pid;

# Load dynamic modules. See /usr/share/doc/nginx/README.dynamic.
include /usr/share/nginx/modules/*.conf;

events {
  worker_connections 1024;
}
```

```
http {
    log_format  main  '$remote_addr - $remote_user [$time_local] "$request" '
    '$status $body_bytes_sent "$http_referer" ''"$http_user_agent"
    "$http_x_forwarded_for"';

    access_log  /var/log/nginx/access.log  main;

    sendfile            on;
    tcp_nopush          on;
    tcp_nodelay         on;
    keepalive_timeout   65;
    types_hash_max_size 4096;

    include             /etc/nginx/mime.types;
    default_type        application/octet-stream;

    include /etc/nginx/conf.d/*.conf;
[...]
    server {
        listen 80;
        listen [::]:80;
        server_name lb-webserver.pro www.lb-webserver.pro;
        return 301 https://$server_name$request_uri;
    }

# Settings for a TLS enabled server.
    server {
        listen 8501 ssl;
        server_name lb-webserver.pro www.lb-webserver.pro;
        root            /usr/share/nginx/html;

        ssl_certificate "/etc/pki/nginx/lb-webserver_pro_combined.crt";
        ssl_certificate_key "/etc/pki/nginx/private/private.key";
        ssl_session_cache shared:SSL:1m;
        ssl_session_timeout  10m;
        ssl_prefer_server_ciphers on;
```

```
    include /etc/nginx/default.d/*.conf;

    error_page 404 /404.html;
        location = /40x.html {
    }

    error_page 500 502 503 504 /50x.html;
        location = /50x.html {
    }

    location / {
        include proxy_params;
        proxy_pass http://streamlit_calc:8501;
        proxy_http_version 1.1;
        proxy_set_header Upgrade $http_upgrade;
        proxy_set_header Connection "upgrade";
        proxy_set_header Host $host;
        proxy_set_header X-Real-IP $remote_addr;
        proxy_set_header X-Forwarded-For $proxy_add_x_forwarded_for;
        proxy_set_header X-Forwarded-Proto $scheme;
    }

  }
[...]
}
```

Placing SSL Certificates and Flask Credentials

Place your SSL certificates into the appropriate directories as described previously in the section "Set Up the SSL Certificates" in Chapter 4.

- SSL certificate

  - ~/Documents/GitHub/deploy-secure-ds-apps-book/config/
    lb-webserver_pro_combined.crt

- SSL private key

  - ~/Documents/GitHub/deploy-secure-ds-apps-book/config/
    private/private.key

Place your Flask authentication file as described in the section "Deploy a Flask App for API Calls with Authentication" in Chapter 5.

- Flask Auth file

  - `~/Documents/GitHub/deploy-secure-ds-apps-book/flask/config/auth_flask.json`

Updating DNS Settings on Namecheap

To configure your domain to point to your Google Cloud VM's external IP:

1. Log in to your Namecheap account.

2. From the left navigation menu, click **Domain List**.

3. Locate your domain and click **MANAGE**.

4. Go to the **Advanced DNS** tab.

5. Adjust the settings by adding or modifying an `A Record` as follows:

   - **Type:** `A Record`

   - **Host:** `@`

   - **Value:** Enter your static external IP (the one you reserved and attached to your VM in the section "Assigning a Static IP and Connecting to the VM via SSH").

   - **TTL:** Set to `1 min`.

Installing Docker Extension for VS Code

For an enhanced development experience, install the official Docker extension by Microsoft for Visual Studio Code:

- Follow the instructions provided previously in the section "Deploy an HTTPS Streamlit App with Domain Name" in Chapter 4.

Updating Google Cloud Firewall Rules

Previously, you configured the firewall to allow ingress traffic on port 8501 (section "Firewall Rules with Network Tagging (Allow 8501 Ingress)"). Now, extend this rule to include ports 8502 and 8504:

1. Navigate to your Google Cloud Console and open **VPC Network ➤ Firewall**.

2. Locate and select the existing firewall rule `allow-rule-webserver`.

3. Edit the **Protocols and Ports** section, adding ports 8502 and 8504 to the existing entry. The final entry should look like `8501,8502,8504`.

4. Save the updated firewall rule.

Deploying the Applications with Docker Compose

With all configurations in place, deploy your applications using Docker Compose; see Code-block 7-19.

Code-block 7-19. Deploy the applications running docker-compose

```
cd ~/Documents/GitHub/deploy-secure-ds-apps-book
docker-compose up -d --build
```

If, after stopping and restarting Docker Compose, you encounter permission errors (particularly with Jenkins files created after initial deployment), rerun the following commands to ensure correct permissions; see Code-block 7-20.

Code-block 7-20. Rerun the permissions

```
cd ~/Documents/GitHub/deploy-secure-ds-apps-book/jenkins
sudo chown -R $(id -u jenkins):jenkins_shared var/jenkins_home
sudo chmod -R 775 var/jenkins_home
```

Your infrastructure should now be fully deployed with secure SSL configurations, and all your applications should be accessible via their respective ports.

Level 3: Deploy with Subdomains

Deploying your applications with subdomains enhances the clarity and accessibility of your infrastructure, enabling each application to have its dedicated URL. Follow these steps carefully to set up your deployment using subdomains.

Update docker-compose.yaml and nginx.conf

Replace your current configuration files with versions configured for subdomain access. Use the provided code-blocks for reference:

- Replace your current `docker-compose.yaml` with the one provided in Code-block 7-22.

- Replace your current `nginx.conf` with the one provided in Code-block 7-23.

Ensure Proper Permissions

To avoid any permission-related issues, rerun these commands to confirm that permissions are correctly set for Jenkins from Code-block 7-20.

Configure DNS Records on Namecheap

Adjust your DNS settings in Namecheap to associate your subdomains with your static external IP:

1. Log in to Namecheap and go to **Domain List**.

2. Find your domain, click **MANAGE**, then navigate to **Advanced DNS**.

3. Add or adjust the following records:

   - One `A Record` with **Host**: @ and **Value**: `<External IP>`

   - Three additional `A Records`:

     - **Host**: `streamlit`, **Value**: `<External IP>`

- **Host**: flask, **Value**: <External IP>

- **Host**: jenkins, **Value**: <External IP>

- Set **TTL** to 1 min for all records.

Replace <External IP> with your VM's static IP address previously configured. Refer to Figure 7-11.

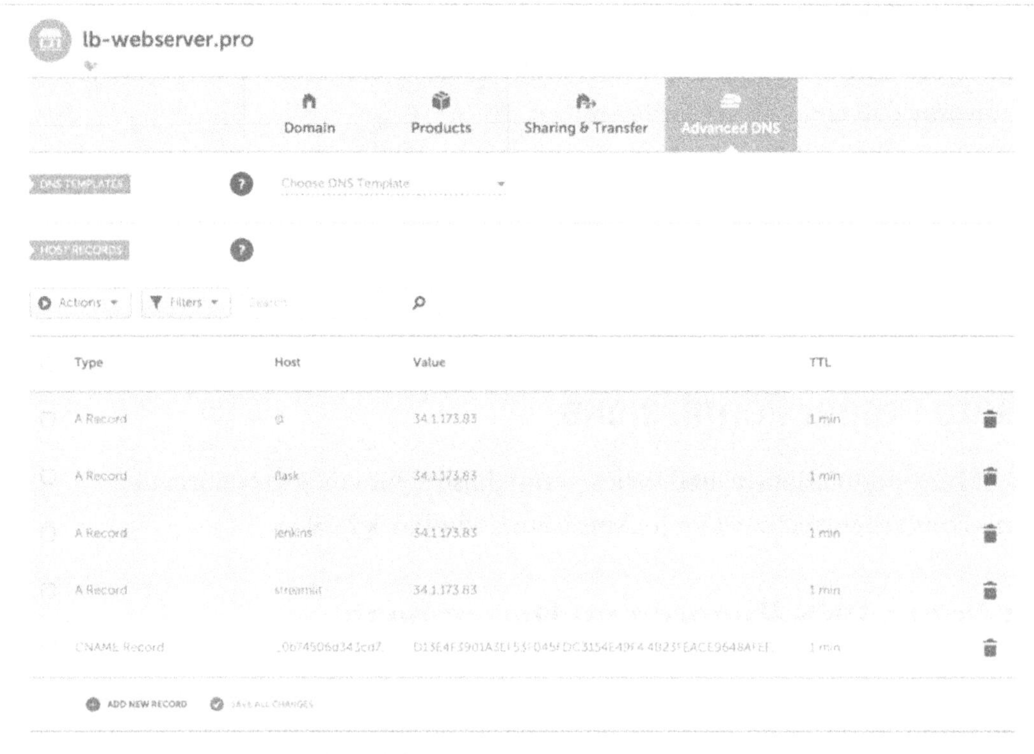

Figure 7-11. *Adjusting Namecheap to work with subdomains for Google Cloud*

Update Firewall Rules

For enhanced security, update your Google Cloud Firewall rules:

- Allow traffic on standard web ports:

- **TCP port 80** (HTTP)

- **TCP port 443** (HTTPS)

- Remove previously opened ports (8501, 8502, and 8504) as these will no longer be necessary.

Deploying Applications with Docker Compose

Deploy your updated configurations by running Docker Compose; refer to Code-block 7-21.

Code-block 7-21. Docker Compose

```
cd ~/Documents/GitHub/deploy-secure-ds-apps-book
sudo docker-compose down
sudo docker-compose up -d --build
```

Your infrastructure is now configured to utilize subdomains, making each application securely and conveniently accessible via

- `https://streamlit.lb-webserver.pro`

- `https://flask.lb-webserver.pro`

- `https://jenkins.lb-webserver.pro`

Code-block 7-22. docker-compose.yaml file with subdomains

```
version: "3.7"

services:
 streamlit_calc:
   build:
     context: streamlit_calc
   ports:
     - "8501"
   restart: always

 flask:
   build:
     context: flask
   ports:
     - "8502"
   volumes:
     - ./flask/logs:/app/logs
   restart: always
```

```yaml
jenkins:
  build:
    context: jenkins
  ports:
    - "8080"
  user: "${JENKINS_UID}:${JENKINS_GID}"
  environment:
    - JENKINS_UID=${JENKINS_UID}
    - JENKINS_GID=${JENKINS_GID}
    - AWS_CONFIG_FILE=/app/.aws/config
  volumes:
    - ./streamlit_calc/data:/app/data
    - ./jenkins/.aws:/app/.aws
    - ./jenkins/var/jenkins_home:/var/jenkins_home # Remember to sudo
      chown -R 1001:1001 jenkins/var/jenkins_home/ and sudo chmod 775
      jenkins/var/jenkins_home/ (so that the volume can be used by
      docker user)
  restart: always

nginx:
  image: nginx:latest
  volumes:
    - ./config/nginx.conf:/etc/nginx/nginx.conf:ro
    - ./config/lb-webserver_pro_combined.crt:/etc/pki/nginx/lb-webserver_
      pro_combined.crt:ro
    - ./config/private/private.key:/etc/pki/nginx/private/private.key:ro
    - ./config/proxy_params:/etc/nginx/proxy_params:ro
  depends_on:
    - streamlit_calc
    - jenkins
    - flask
  ports:
    - "80:80"
    - "443:443"
  restart: always
```

Code-block 7-23. nginx.conf file with subdomains

```
user nginx;
worker_processes auto;
error_log /var/log/nginx/error.log;
pid /run/nginx.pid;

events {
   worker_connections 1024;
}

http {
   log_format  main  '$remote_addr - $remote_user [$time_local]
   "$request" '
                      '$status $body_bytes_sent "$http_referer" '
                      '"$http_user_agent" "$http_x_forwarded_for"';

   access_log  /var/log/nginx/access.log  main;

   sendfile            on;
   tcp_nopush          on;
   tcp_nodelay         on;
   keepalive_timeout   65;
   types_hash_max_size 4096;

   include             /etc/nginx/mime.types;
   default_type        application/octet-stream;

   server {
       listen 80;
       listen [::]:80;
       server_name streamlit.lb-webserver.pro flask.lb-webserver.pro
       jenkins.lb-webserver.pro;
       return 301 https://$host$request_uri;
   }
```

```
# Settings for a TLS enabled server.
   # Streamlit subdomain
   server {
       listen 443 ssl;
       server_name streamlit.lb-webserver.pro;

       ssl_certificate "/etc/pki/nginx/lb-webserver_pro_combined.crt";
       ssl_certificate_key "/etc/pki/nginx/private/private.key";
       ssl_session_cache shared:SSL:1m;
       ssl_session_timeout  10m;
       ssl_prefer_server_ciphers on;

       include /etc/nginx/default.d/*.conf;

       error_page 404 /404.html;
           location = /40x.html {
       }

       error_page 500 502 503 504 /50x.html;
           location = /50x.html {
       }

       location / {
           include proxy_params;
           proxy_pass http://streamlit_calc:8501;
           proxy_http_version 1.1;
           proxy_set_header Upgrade $http_upgrade;
           proxy_set_header Connection "upgrade";
           proxy_set_header Host $host;
           proxy_set_header X-Real-IP $remote_addr;
           proxy_set_header X-Forwarded-For $proxy_add_x_forwarded_for;
           proxy_set_header X-Forwarded-Proto $scheme;
       }

   }

[...]
}
```

Summary

In this chapter, you replicated your AWS deployment architecture onto Google Cloud Platform (GCP). Specifically, you

- Created a GCP account and project and configured IAM roles and permissions

  - GCP IAM and Admin documentation: `https://cloud.google.com/iam/docs`

- Launched a virtual machine (VM) using an automated startup script to install necessary software (Python, Docker, Docker Compose, Git)

  - GCP Compute Engine documentation: `https://cloud.google.com/compute/docs`

- Assigned a static IP address to your VM and connected securely via SSH using VS Code

- Deployed a simple Streamlit application on port 8501, configuring firewall rules to enable external access

- Cloned your project repository from GitHub, configured user permissions, and deployed all applications (Streamlit, Flask, Jenkins) with SSL certificates from your repository

- Updated DNS settings on Namecheap to point your domain to your GCP VM's external IP

- Enhanced your deployment structure by configuring subdomains for each application (`streamlit.lb-webserver.pro`, `flask.lb-webserver.pro`, `jenkins.lb-webserver.pro`), updating Docker Compose, Nginx configuration, DNS records, and firewall rules accordingly

  - GCP Firewall Rules documentation: `https://cloud.google.com/firewall/docs/firewalls`

In the next chapter, you'll advance your GCP deployment by implementing autoscaling and load balancing across global regions, configuring instance templates, setting up instance groups, modifying firewall rules, and optimizing load balancers to handle HTTPS traffic efficiently.

Advanced Deployment in GCP: Autoscaling and Load Balancing Across Global Regions

In the previous chapter, you successfully replicated your comprehensive AWS-based deployment infrastructure on Google Cloud Platform (GCP). You created a GCP account and project, configured IAM roles, launched virtual machines (VMs), deployed applications progressively (Streamlit, Flask, Jenkins), and secured your setup using SSL certificates and subdomains.

In this chapter, we build upon the foundation setup of Chapter 7 by introducing advanced deployment features available on GCP. Specifically, we'll implement autoscaling capabilities and global load balancing to ensure your infrastructure can dynamically respond to varying demands, optimizing performance and reliability. You will learn how to create VM images, configure instance templates, establish managed instance groups with autoscaling, modify firewall rules to allow HTTPS traffic, and set up a global HTTPS load balancer secured with SSL certificates and subdomain support.

Creating an Image from the Existing VM's Disk

In this step, you'll create an image from the disk of your existing virtual machine. This image will allow you to easily replicate your current setup on new VM instances without needing to manually reconfigure everything.

© Lucas H. Benevides e Braga 2025
L.H.B.e. Braga, *Deploying Secure Data Science Applications in the Cloud*,
https://doi.org/10.1007/979-8-8688-1715-1_8

Step-by-Step Instructions

1. **Modify VM disk deletion rule**

   - Go to **Compute Engine ➤ VM Instances** in the Google Cloud Console.

   - Select your existing VM instance, then click **EDIT**.

   - Scroll down to the **Storage** section.

   - Under **Deletion rule**, select **Keep disk** (this ensures the disk remains available even after the VM instance is deleted).

   - Click **SAVE** at the bottom of the page.

   See Figure 8-1.

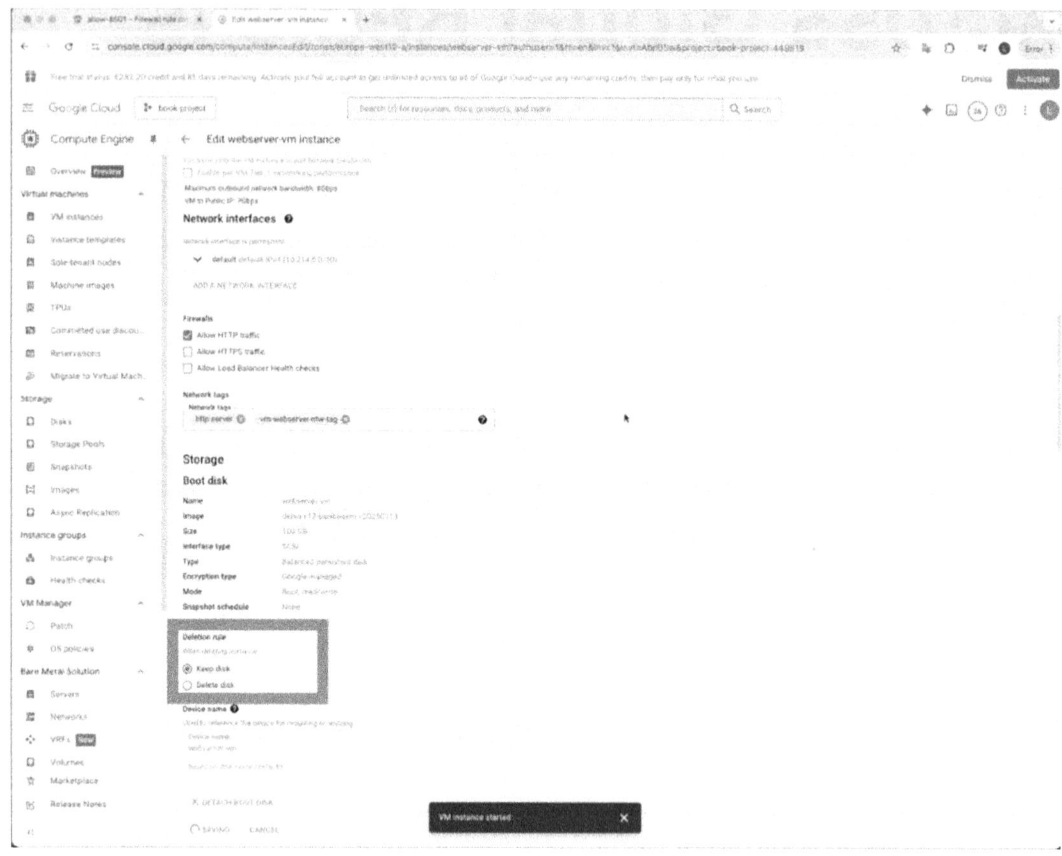

Figure 8-1. *Keep disk*

2. **Stop and delete the VM instance**

   - Return to **Compute Engine ➤ VM Instances**.

   - Select your VM instance and click **STOP**.

   - After the instance has successfully stopped, click **DELETE** to remove it completely (the disk will remain due to the setting configured previously).

3. **Create an image from the existing disk**

   - Navigate to **Compute Engine ➤ Storage ➤ Images**.

   - Click **CREATE IMAGE**.

   - Configure the new image as follows:

     - **Name:** Enter a clear and descriptive name, such as `webserver-image-with-subdomains`.

     - **Source:** Select **Disk**.

     - **Source disk:** Choose the disk previously attached to your deleted VM instance (e.g., `webserver-vm`).

     - **Location:** Choose **Multi-regional**.

     - **Select location:** Select `eu` (for multiple regions within the European Union).

   - Leave the remaining default settings as is.

   - At the bottom, click **CREATE** and wait until the image creation process completes.

See Figure 8-2.

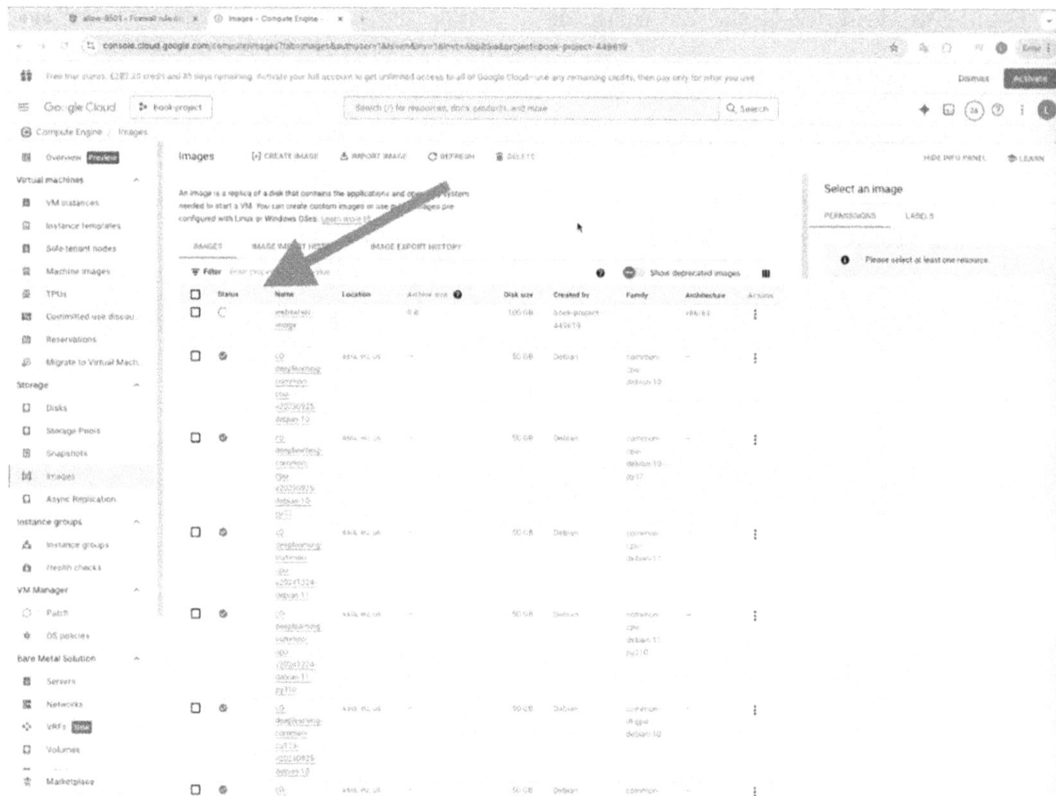

Figure 8-2. *Create image*

Your custom image is now ready and can be used for creating identical VM instances efficiently across multiple regions.

Configure an Instance Template with a Custom Image and Network Tags

Now that you have successfully created your custom image, the next step is to use this image to configure an instance template. This template will be used later to easily deploy multiple VM instances across various regions.

Step-by-Step Instructions

1. **Navigate to Instance Templates**

   - In the Google Cloud Console, select **Compute Engine ➤ Virtual Machines**.

   - Click **Instance templates**, then select **CREATE INSTANCE TEMPLATE**.

2. **Configure the Instance template**

   - **Name:** Enter a descriptive name, such as `webserver-template-with-subdomains`.

   - **Location:** Choose **Global**.

   - **Machine configuration**

     - Set **Machine type** to **Standard**.

     - Select the machine type as **e2-standard-4** (4 vCPUs, 2 cores, and 16 GB memory per instance).

3. **Set up the custom boot disk**

   - Scroll down to the **Boot disk** section and click **CHANGE**.

   - Navigate to the **Custom images** tab.

   - Choose the image you previously created (e.g., `webserver-image-with-subdomains`, the disk size will automatically reflect the size from your original VM, e.g., 100 GB).

   - Click **SELECT** to confirm.

4. **Configure network tags**

   - Scroll down to the **Networking** section.

   - In the **Network tags** field, enter the tag you previously configured, such as `vm-network-tag`. (Ensure this matches the tag associated with your existing firewall rules. If uncertain, revisit the firewall rule `allow-rule-webserver` to confirm the correct network tag.)

5. **Finalize and create the template**

- Review your settings, leaving all other defaults unchanged.

- At the bottom of the page, click **CREATE**.

Your instance template with the custom image and network tags is now ready for deployment, simplifying the creation of identical, scalable VM instances.

Set Up an Instance Group with Autoscaling Based on Load Balancer Utilization and Health Checks

In this step, you will create two instance groups (one for Europe and another for the United States) using the instance template created previously. Both instance groups will utilize autoscaling features to dynamically manage the number of VM instances based on HTTP load balancer utilization. Additionally, you'll configure health checks to continuously monitor instance health and ensure optimal availability.

Creating the Europe Instance Group

1. **Navigate to Instance Groups**

- Go to **Compute Engine ➤ Instance groups**.

- Click **CREATE INSTANCE GROUP**.

See Figure 8-3.

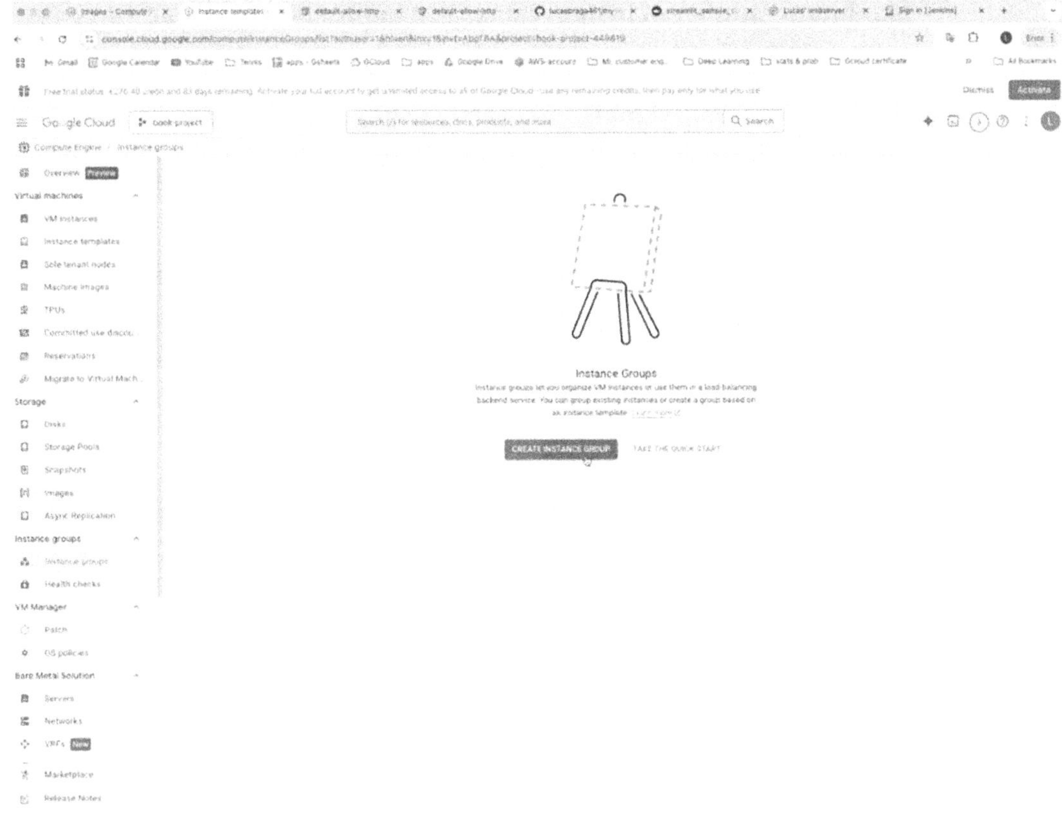

Figure 8-3. *Create instance group*

2. **Configure Instance Group (Europe)**

   - **Name:** Enter a descriptive name, such as `webserver-instance-group-europe`.

   - **Instance template:** Select the previously created template, `webserver-template-with-subdomains`.

   - **Location:** Choose **Multiple zones** and select a European zone, for example, `europe-west10 (Berlin)`.

3. **Set up Autoscaling**; see Figure 8-4.

   - **Autoscaling mode:** Select **On: add and remove instances to the group**.

   - **Minimum number of instances:** Set to 1.

- **Maximum number of instances:** Set to 2 (adjust based on your expected traffic and cost considerations).

- Under **Autoscaling signals**, select the existing entry (e.g., CPU utilization) and change it to

  - **Signal type: HTTP load balancing utilization**

  - ***Target HTTP load balancing utilization:*** Set to 80% (once the utilization of the load balancer reaches 80%, then it spins up a new instance to support the increase in traffic/demand)

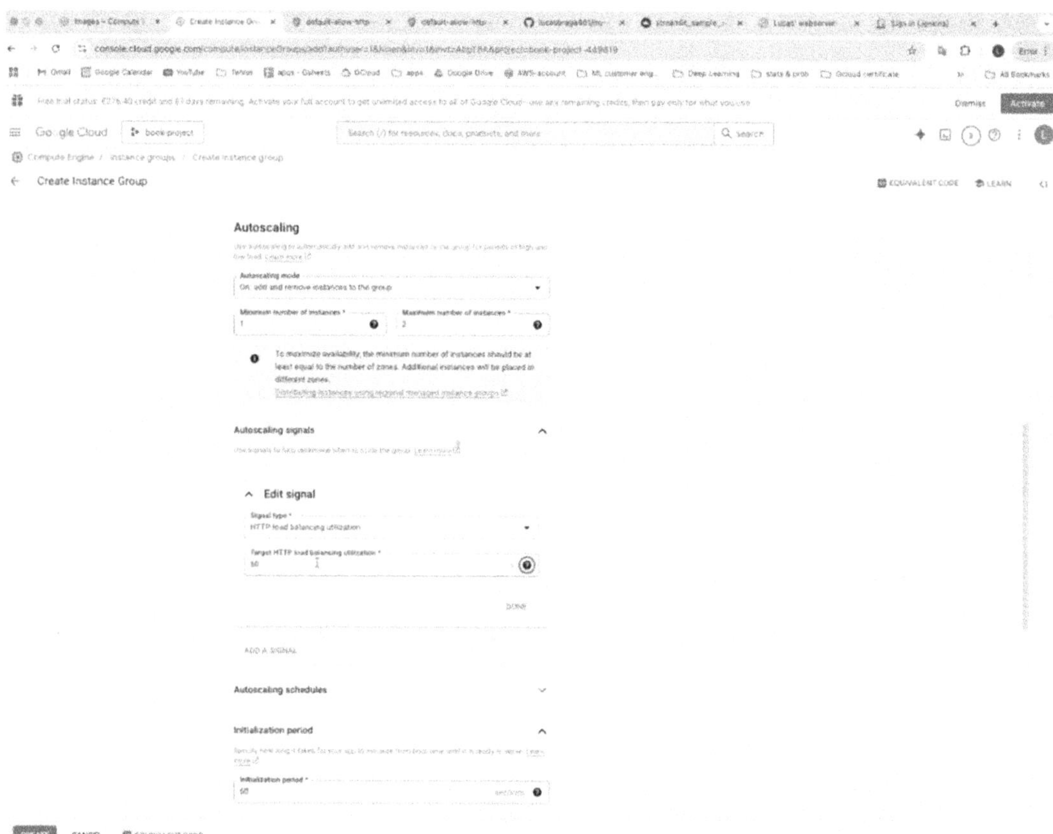

Figure 8-4. *Configure autoscaling*

4. **Configure Autohealing (health check)**; see Figure 8-5.

- In the **VM instance lifecycle** section, click **Health check** and select **CREATE A HEALTH CHECK**.

- In the configuration panel:

  i. **Name:** Enter a descriptive name, for example, `https-health-check`.

  ii. **Scope:** Choose **Global**.

  iii. **Protocol:** Select **TCP**.

  iv. **Port:** Enter 443.

  v. **Health criteria**

     1. **Check interval:** Set to 10 seconds (so that it checks the port every ten seconds).

     2. **Timeout:** Set to 5 seconds (so that it waits five seconds before it calls a timeout error).

     3. **Healthy threshold:** Set to 2 (marks instance healthy after two successful checks).

     4. **Unhealthy threshold:** Set to 3 (marks instance unhealthy after three failed checks).

- Click **SAVE**.

- Set the **Initial delay** to 60 seconds (allows instances adequate time to initialize).

5. **Finalize and create the Europe instance group**

- Verify your settings and click **CREATE** at the bottom of the page.

- A message will appear stating "Autoscaling configuration is not complete," reminding you to later configure the load balancer. Click ***CONFIRM***.

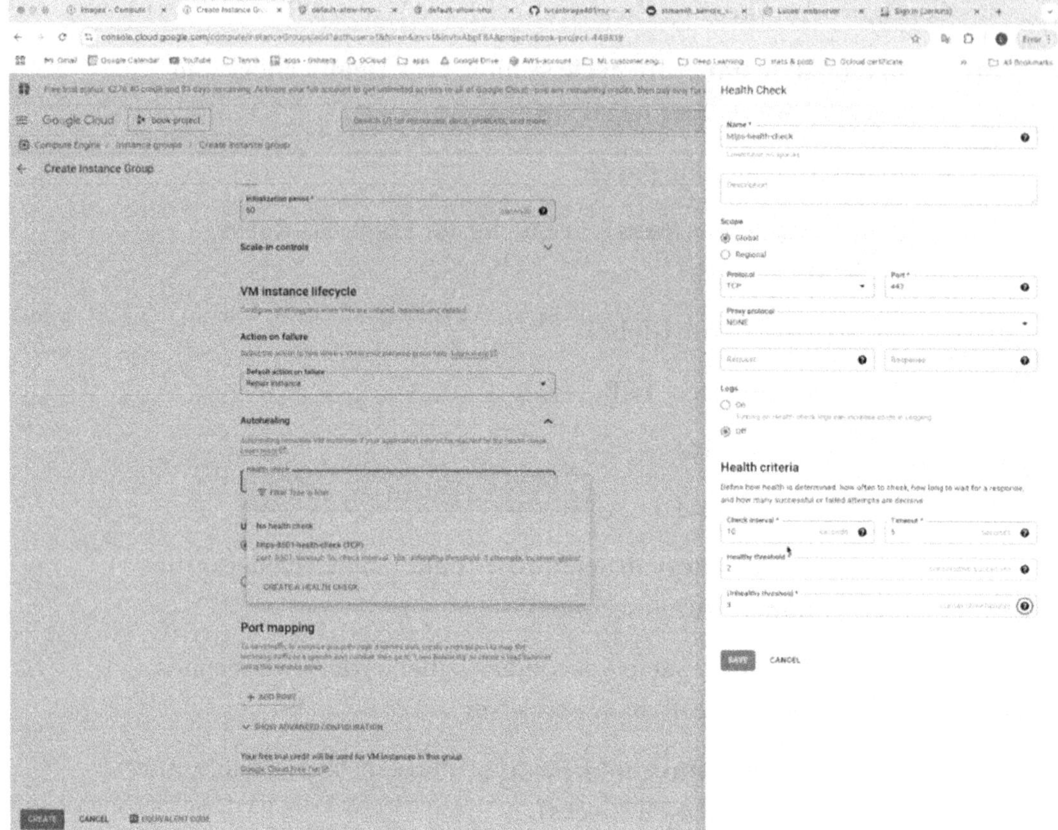

Figure 8-5. *Create a health check.*

Creating the US Instance Group

Repeat a similar process to create the US instance group:

1. **Navigate to Instance Groups**

   - Return to **Compute Engine ➤ Instance groups** and click **CREATE INSTANCE GROUP**.

2. **Configure Instance Group (US)**

   - **Name:** Use a descriptive name such as `webserver-instance-group-us`.

   - **Instance template:** Again select `webserver-template-with-subdomains`.

   - **Location:** Choose **Multiple zones** and select a US region, for example, `us-east4 (Northern Virginia)`.

3. **Autoscaling settings**

- Configure identically to the European instance group:

- Autoscaling mode: **On: add and remove instances to the group**

- Minimum instances: 1

- Maximum instances: 2

- Autoscaling signals:

    - **Signal type:** HTTP load balancing utilization

    - **Target utilization:** 80%

4. **Configure Autohealing**

- Select the previously created **https-health-check**.

- Set the **Initial delay** to 60 seconds.

5. **Finalize and create the US instance group**

- Click **CREATE** and then **CONFIRM** when prompted.

Refer to Figure 8-6 for reference during instance group creation.

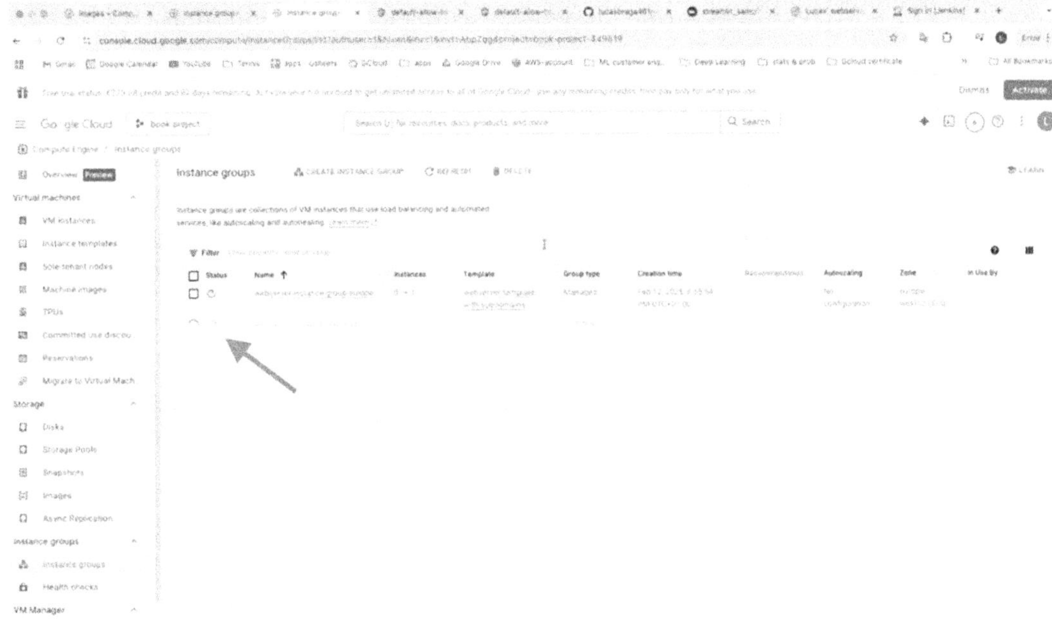

Figure 8-6. *Waiting for the US instance group to be created*

Once both instance groups (Europe and United States) have been successfully created and are operational, proceed to the next section to configure the global load balancer.

Modify Firewall Rules to Permit HTTPS Ingress Traffic (Port 443) for Subdomain Compatibility

To enable your applications to securely handle HTTPS traffic via subdomains, you must update your Google Cloud Firewall rules to allow ingress on port 443. Follow these detailed steps:

Step-by-Step Instructions

1. **Navigate to Firewall settings**

   - In the Google Cloud Console, select **VPC Network ➤ Firewall**.

2. **Modify existing firewall rule (or create a new one)**

   - Locate the existing firewall rule created earlier, such as `allow-rule-webserver`. Alternatively, you may choose to create a new firewall rule (e.g., `webserver-firewall`) for improved clarity. If you create a new rule, remember to delete the previous one afterward to maintain organization.

3. **Configure firewall rule**

   Adjust the following settings in your firewall rule:

   - **Direction of traffic:** Set to **Ingress**.

   - **Action on match:** Choose **Allow**.

   - **Targets:** Select **Specified target tags**.

   - **Target tags:** Enter the tag you previously used (`vm-network-tag`). This tag must match the network tag defined during instance template creation (see section "Configure an Instance Template with a Custom Image and Network Tags").

     (This ensures the firewall rule correctly identifies which instances it applies to.)

- **Source IPv4 ranges:** Specify the IP addresses permitted to access your infrastructure. Options include

  - To limit access only to your IP:

    - Find your IP at whatismyipaddress.com, and add it with /32 at the end, for example, `87.154.215.35/32`. That could also be the IP from a corporate VPN that is mutual among multiple users.

  - To allow a specific IP range:

    - Adjust the IP and suffix accordingly (see the note below).

  - To open access globally (not recommended unless necessary):

    - Use `0.0.0.0/0`.

- **Protocols and ports:** Select **Specified protocols and ports**.

  - Enter **TCP:** 443 for secure HTTPS traffic with subdomains.

  - Optionally, add **TCP:** 22 if you require SSH access via this rule (if more than one port, add them separated by a comma, e.g., `443, 22`, or handle it through another dedicated rule).

4. **Finalize and save the firewall rule**

   - Click **CREATE** or **SAVE** at the bottom of the configuration panel. See Figure 8-7.

Note   Understanding IP Range Suffixes (CIDR Notation)

What's with the suffix /32 at the end of an IP range for the Firewall rule?

Using /32 means that the rule applies only to that exact IP address. But for each number that you decrease, the range doubles. So

- /31 -> 2 IP addresses

- /30 -> 4 IP addresses

- /29 -> 8 IP addresses

- ...

- /24 -> 256 IP addresses

- ...

- /16 -> 65,536 IP addresses

In that case if the rule is applied to 192.168.1.1**/32**, it will only allow traffic to that exact 192.168.1.1, but if the rule is applied instead to

- 192.168.1.0**/24**, then it will allow traffic from all IP addresses in the range of 192.168.1.0 to 192.168.1.255; that's 256 IPs.

- 192.168.0.0**/21**, then it will allow traffic from all IP addresses in the range of 192.168.1.0 to 192.168.7.255; that's 2,048 IPs.

- 192.168.0.0**/16**, then it will allow traffic from all IP addresses in the range of 192.168.0.0 to 192.168.255.255; that's 65,536 IPs.

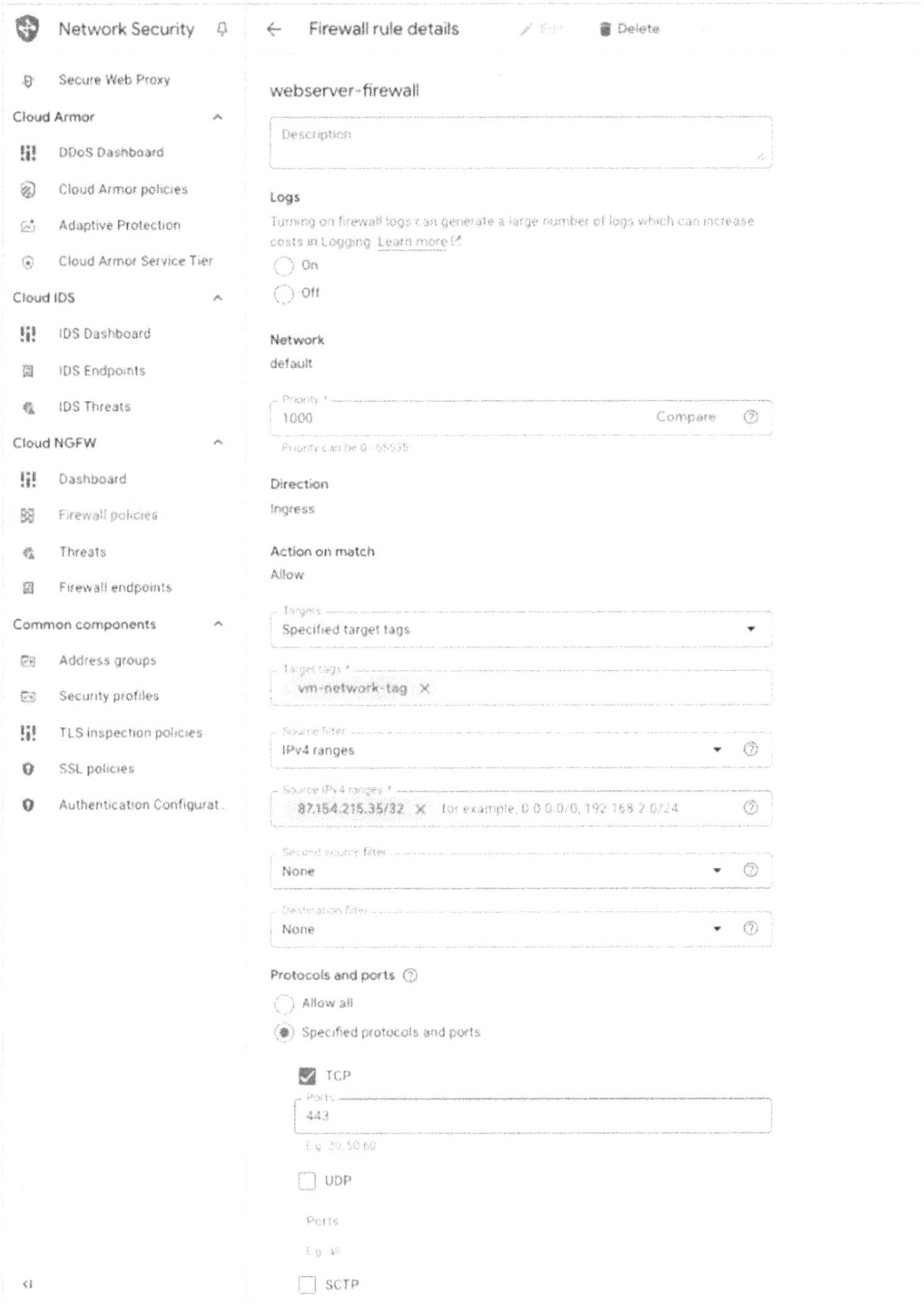

Figure 8-7. *Adjust the firewall rule*

Set Up a Global Load Balancer with SSL, HTTPS, and Subdomain Support

In this step, you will set up a global load balancer configured to securely handle HTTPS traffic using SSL certificates and subdomains. First, you'll need to upload your existing SSL certificates to Google Cloud, making them available for use with the load balancer.

Uploading SSL Certificates to Google Cloud

Before creating the load balancer, you need to upload your SSL certificates to your Google Cloud project:

Step-by-Step Instructions

1. **Open Google Cloud Shell Terminal**

   - In the Google Cloud Console, click the small terminal icon located in the top-right corner to open the Google Cloud Shell Terminal.

   - Once opened, click **Open Editor** (VS Code-like environment provided by Google Cloud). See Figure 8-8.

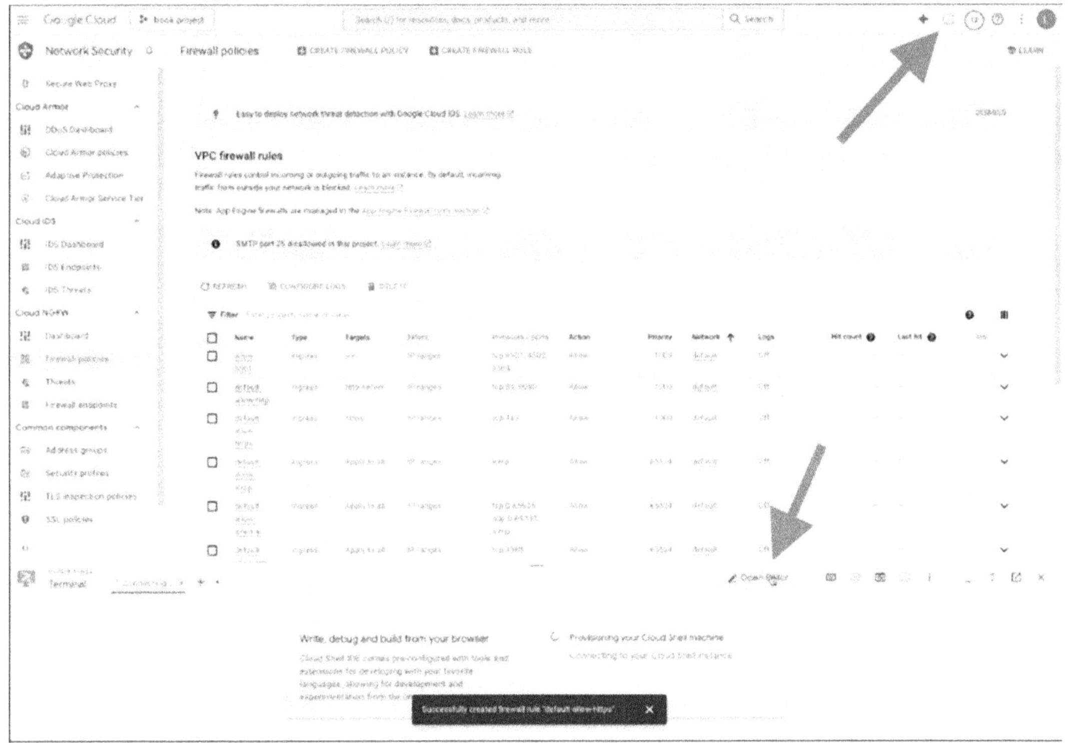

Figure 8-8. *Open shell terminal and GCloud VS Code*

2. **Upload SSL certificates**

- In your Cloud Shell Editor, navigate to your home folder (~/) and create a new directory named `config`.

- Within this `config` directory, upload your combined SSL certificate file (previously named `lb-webserver_pro_combined.crt`).

- Inside the `config` folder, create another directory named `private`, and upload your private key file (`private.key`) into this folder.

Your folder structure should look like this from Code-block 8-1.

Code-block 8-1. Folder structure with SSL certificate files

```
~/config/lb-webserver_pro_combined.crt
~/config/private/private.key
```

3. **Create SSL certificates resource in Google Cloud**

- With the files uploaded, run the following command in your
  Google Cloud Shell Terminal to create an SSL certificate resource
  within your Google Cloud project. See Code-block 8-2.

Code-block 8-2. Send certificates to your GCloud project

```
gcloud compute ssl-certificates create lb-webserver-cert \
    --certificate=config/lb-webserver_pro_combined.crt \
    --private-key=config/private/private.key \
    --global
```

Refer to Figure 8-9 for reference on this process.

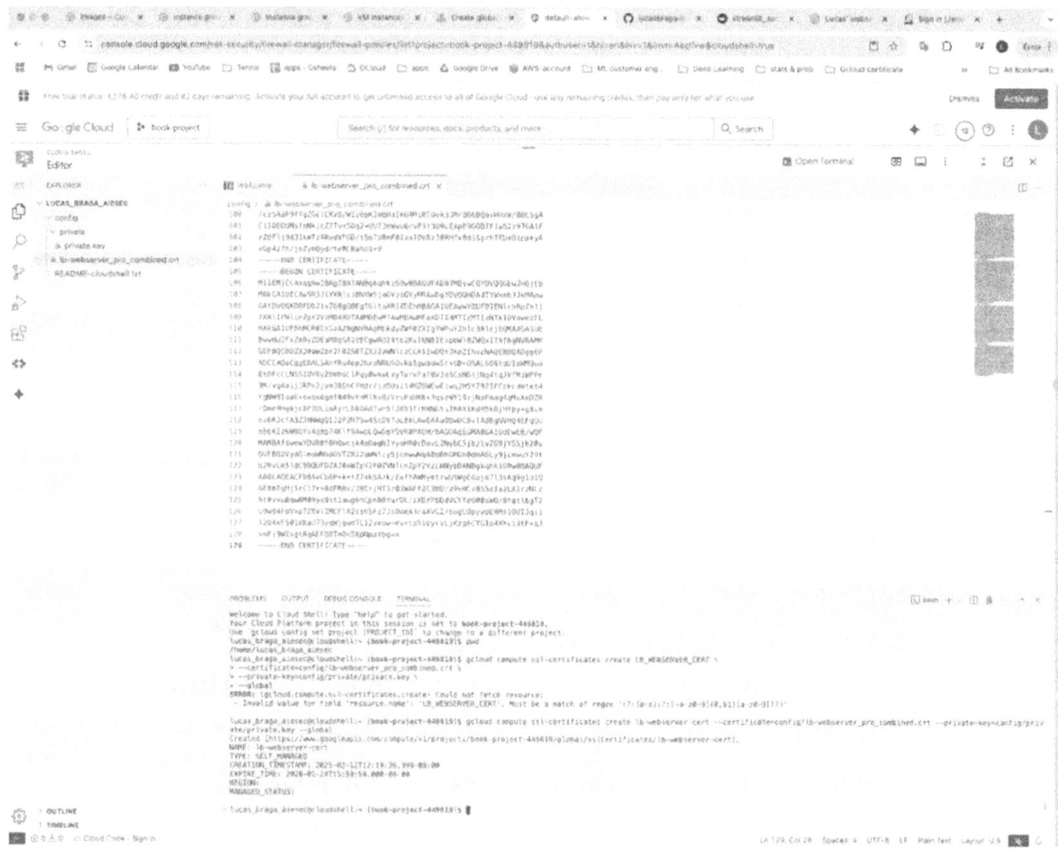

Figure 8-9. *Uploading the SSL certificates*

After executing this command, your SSL certificates will be stored securely in your Google Cloud project and ready for use in the next steps, where you will create the global HTTPS load balancer.

Creating the Load Balancer

With your SSL certificates successfully uploaded, the next step is to create and configure a global HTTPS load balancer in Google Cloud to distribute incoming requests effectively to your US and Europe instance groups.

Step-by-Step Instructions

1. **Navigate to Load Balancing**

    - In the Google Cloud Console, select **Network Services ➤ Load Balancing** (or use the top search bar to search for "Load Balancing").

    - Click **CREATE LOAD BALANCER** to begin.

2. **Configure load balancer (initial setup)**

    - **Type of load balancer:** Select **Application Load Balancer (HTTP/HTTPS)**; click **NEXT**.

    - **Public facing or internal:** Choose **Public facing (external)**; click **NEXT**.

    - **Global or single region deployment:** Choose **Best for global workloads**; click **NEXT**.

    - **Load balancer generation:** Select **Global external Application Load Balancer**, click **NEXT**, then click **CONFIGURE**.

    See Figure 8-10.

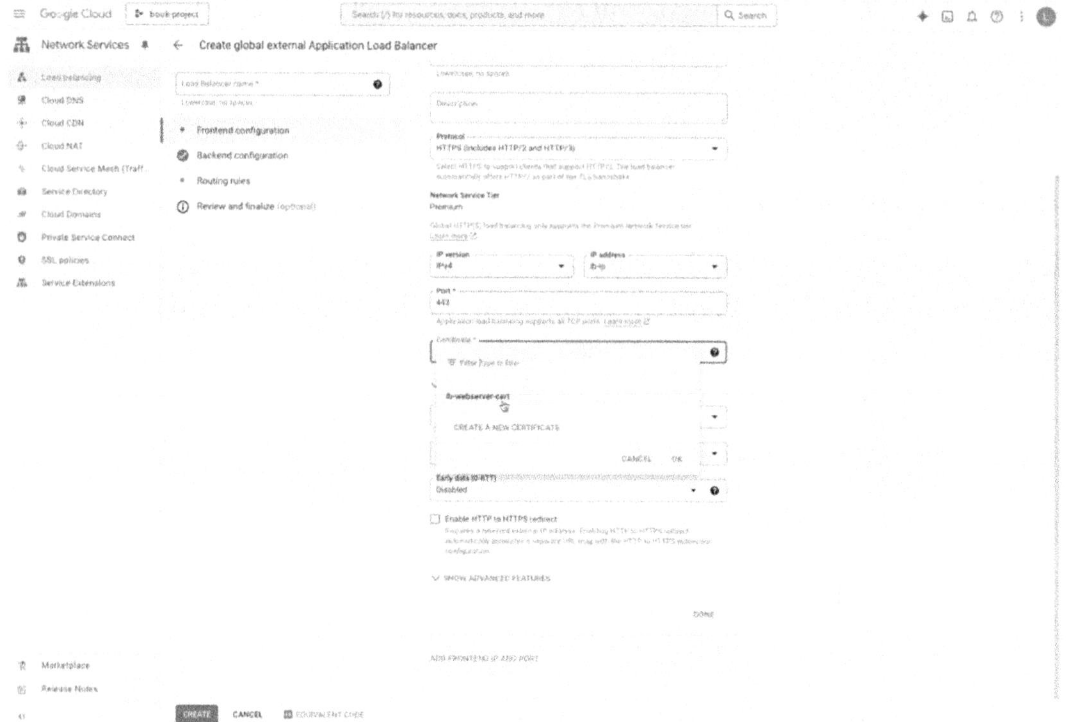

Figure 8-10. *Start creating the load balancer.*

3. **Frontend configuration**

   Fill out the frontend details as follows:

   - **Name:** Enter a descriptive name such as `https-lb-webserver`.

   - **Protocol:** Select **HTTPS (includes HTTP/2 and HTTP/3).**

   - **IP version:** Choose **IPv4.**

   - **IP address:** Click and select **CREATE IP ADDRESS** from the drop-down.

     - Enter a descriptive name, for example, `lb-ip`, then click **RESERVE.**

   - **Port:** Enter 443.

   - **Certificate:** Click and select the SSL certificate `lb-webserver-cert` uploaded previously.

   - Click **DONE.**

4.   **Backend configuration**

Under **Backend services & backend buckets**, click and select
CREATE A BACKEND SERVICE.

Configure as follows:

- **Name:** Enter a clear name such as `https-lb-backend-webserver`.

- **Backend type:** Choose **Instance group**.

- **Protocol:** Set to **HTTPS**.

- **Named port:** Enter a descriptive name, for example, `https-port`.

- **Timeout:** Set to 30 seconds.

- **IP address selection policy:** Choose **Only IPv4**.

- **Backends**

  - Click **New backend:**

    - **IP stack type:** Select **IPv4 (single-stack)**.

    - **Instance group:** Select your **Europe instance group**.

    - **Port numbers:** Enter 443.

    - **Balancing mode:** Select **Rate** *(alternative option: Utilization so that the backend service spins up after a certain CPU utilization percentage is achieved instead of number of requests per second which is what Rate is about).*

    - **Maximum RPS:** Set to 50.

    - **Scope:** Choose **per instance**.

    - **Capacity:** Enter 100.

    - **Backend preference level:** Keep as **None**.

  - Click **DONE**.

  - Click **ADD A BACKEND** to add your **US instance group** with identical settings as above, then click **DONE**.

  - **Health check:** Select the previously created **https-health-check**.

- Scroll to **Security**: Leave default values.

    - **Policy name:** Google Cloud may auto-generate a policy name exceeding 63 characters (which is invalid). Shorten this auto-generated policy name if necessary to under 63 characters.

- Click **CREATE** at the bottom of the backend configuration panel.

5. **Finalize load balancer creation**

- Click **OK** in the Backend configuration drop-down.

- Skip the **Routing rules** section (leave defaults as is).

- On the left menu, select **Review and finalize**:

    - Double-check all configurations to ensure accuracy.

- Enter a descriptive **Name** for your load balancer in the field at the top left of the page.

- Click **CREATE** at the bottom left of the page.

See Figure 8-11.

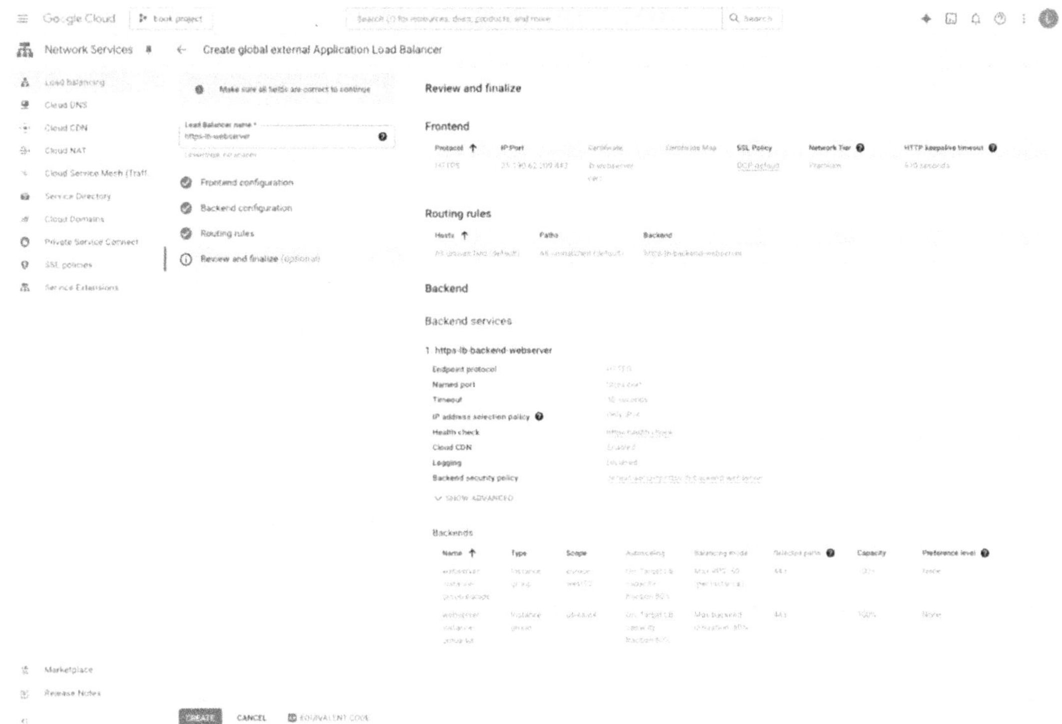

Figure 8-11. *Finish creating the load balancer*

Wait until the load balancer creation process completes successfully. Once done, proceed to the next section.

Updating the DNS Records in Namecheap

With your load balancer successfully created, you now need to update your DNS records on Namecheap, pointing your domain and subdomains to the load balancer's IP address. Follow the steps below:

Step-by-Step Instructions

1. **Obtain Load Balancer IP address**

   - In the Google Cloud Console, navigate to **Network Services ➤ Load Balancing**.

   - Click on your recently created load balancer and note its external IP address (refer to Figure 8-12).

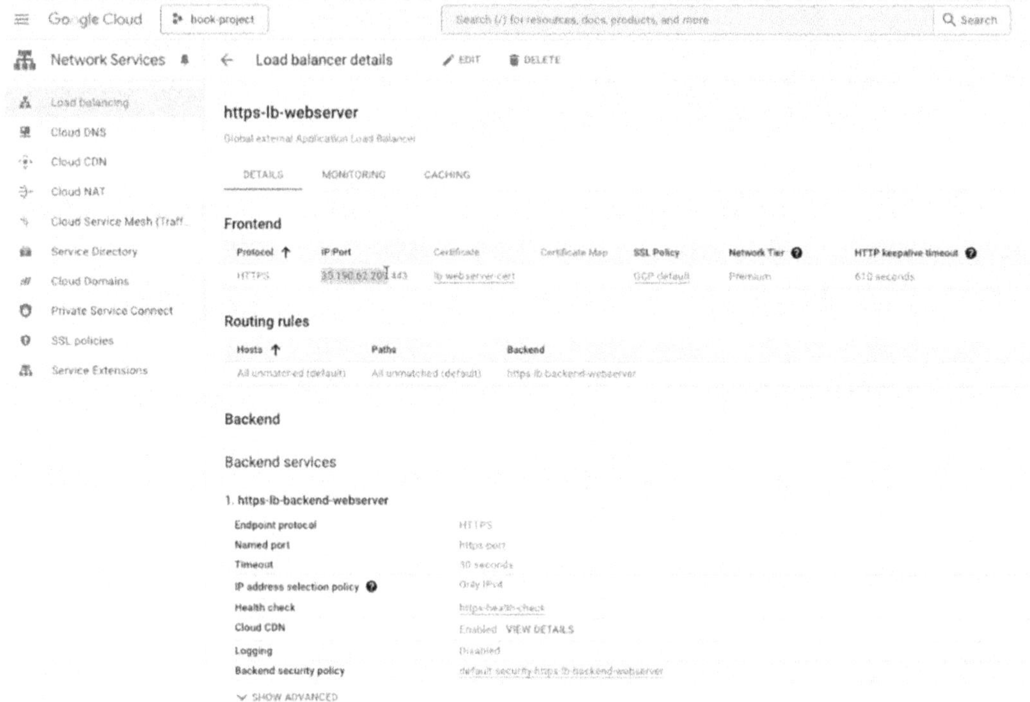

Figure 8-12. *Get the IP from the load balancer*

2. **Update DNS records on Namecheap**

- Log in to your Namecheap account.

- On the left-hand menu, click **Domain List**.

- Locate your domain, click **MANAGE**, and navigate to the
  Advanced DNS tab.

- Update or add DNS records according to the settings shown in
  Table 8-1, visually represented in Figure 8-13.

Table 8-1. *Namecheap IP's setup*

Type	Host	Value	TTL
A Record	@	<Load Balancer's IP>	1 min
A Record	flask	<Load Balancer's IP>	1 min
A Record	jenkins	<Load Balancer's IP>	1 min
A Record	streamlit	<Load Balancer's IP>	1 min
CNAME	_0b7…	D131329…	1 min

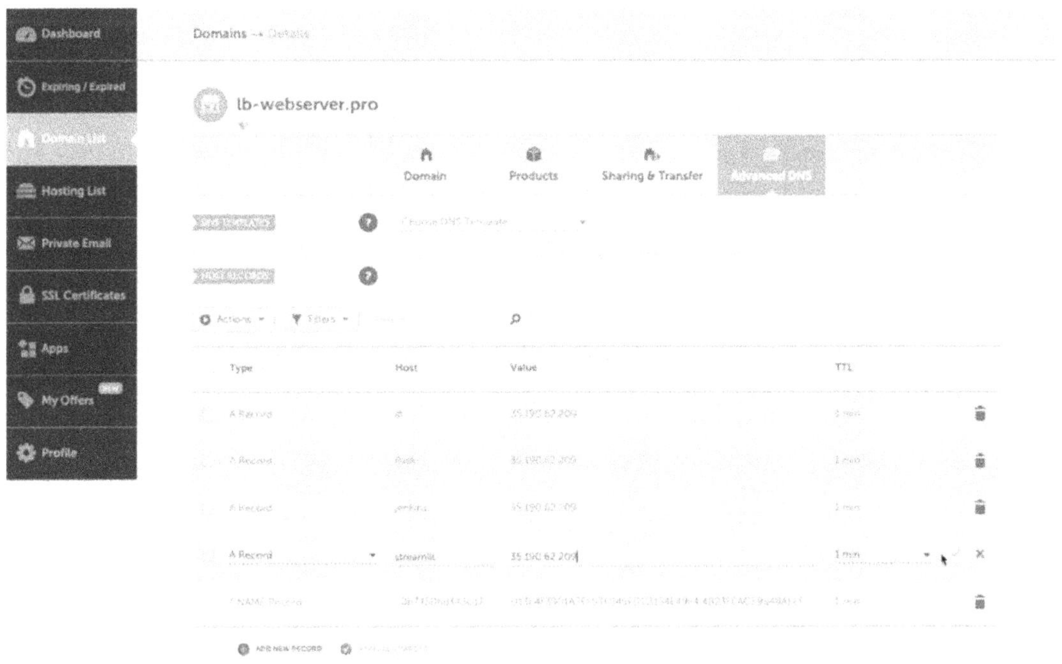

Figure 8-13. *Adjust the IPs in Namecheap*

3. **Verify the changes**

   - After updating the DNS records, wait approximately one or two
     minutes for the changes to propagate.

- Refresh your applications in the browser and confirm they are operational at

  - streamlit.lb-webserver.pro

  - flask.lb-webserver.pro

  - jenkins.lb-webserver.pro

Summary

In this chapter, you enhanced your Google Cloud deployment with advanced scalability and global load balancing features. Specifically, you

- Created a reusable VM image from your existing virtual machine's disk, facilitating rapid, consistent deployments across multiple regions

  - Google Cloud documentation on creating and managing images: `https://cloud.google.com/compute/docs/images`

- Configured an instance template using the custom VM image and applied network tags for seamless firewall management

  - Google Cloud documentation on instance templates: `https://cloud.google.com/compute/docs/instance-templates`

- Established managed instance groups in both Europe and the United States, utilizing autoscaling based on HTTPS load balancer utilization (80%) and health checks to maintain instance reliability

  - Google Cloud documentation on managed instance groups and autoscaling: `https://cloud.google.com/compute/docs/instance-groups`

- Modified GCP firewall rules to permit HTTPS ingress traffic on port 443, ensuring secure subdomain accessibility

- Configured a global HTTPS load balancer with SSL certificates and subdomain support, distributing traffic efficiently across regional instance groups

  - Google Cloud documentation on load balancing: `https:// cloud.google.com/load-balancing/docs/overview`

- Updated DNS records on Namecheap to direct domain and subdomain traffic to your global Load Balancer IP, ensuring proper routing and accessibility

In the next chapter, you'll explore serverless deployment strategies using Google Cloud Run, further enhancing your infrastructure's agility and scalability. You will deploy individual Cloud Run services, configure serverless Network Endpoint Groups (NEGs), set up SSL certificates, establish a secure global load balancer, update DNS records, restrict direct access to Cloud Run URLs, and configure authenticated access.

PART II

Serverless Deployments

## CHAPTER 9

Serverless Deployment with Google Cloud Run

In the previous chapter, you scaled your Google Cloud infrastructure globally using autoscaling managed instance groups and an HTTPS global load balancer. You configured backend health checks, set firewall rules, uploaded your SSL certificates, and connected subdomains to a load-balanced deployment using persistent VM instances.

In this chapter, we will explore deploying applications using a **serverless environment**. Unlike previous deployments using virtual machines (VMs), serverless computing provides an abstraction that lets you run your applications without directly managing server infrastructure. We'll leverage Google Cloud Platform's (GCP) **Cloud Run** and AWS's **Fargate** to handle the infrastructure for us.

Key Concepts and Questions

Before proceeding, let's clarify some foundational concepts:

- **What is serverless computing?**

- **What are stateful, stateless, and ephemeral applications?**

- **How does serverless deployment differ from traditional VM-based deployments we've used earlier?**

Understanding Serverless Computing

Serverless computing refers to cloud-based infrastructure that fully abstracts away server management, enabling developers to build and run applications without worrying about the underlying infrastructure. Providers such as Google Cloud manage provisioning, scaling, and managing the infrastructure required to run your code.

© Lucas H. Benevides e Braga 2025
L.H.B.e. Braga, *Deploying Secure Data Science Applications in the Cloud*,
https://doi.org/10.1007/979-8-8688-1715-1_9

According to Google's official documentation:

> "Serverless computing lets developers focus on their code, rather than managing infrastructure. Serverless automatically scales resources up or down based on usage, making it highly efficient and cost-effective."

For more detailed information, visit the official Google Cloud documentation on *Serverless Computing* (`https://cloud.google.com/serverless`).

Stateful, Stateless, and Ephemeral Applications

- **Stateful applications** store session information or data that persists beyond a single request or interaction.

- **Stateless applications** do not store any session information; each request is independent, making these applications easier to scale.

- **Ephemeral applications** are temporary, short-lived, and often stateless by nature.

A comprehensive explanation of these concepts can be found in Red Hat's guide on *Stateful vs. Stateless Applications* (`https://www.redhat.com/en/topics/cloud-native-apps/stateful-vs-stateless`).

Differences from Previous VM Deployments

Serverless computing differs significantly from VM-based deployments:

- No need to manage underlying infrastructure or operating systems

- Automatic scaling and reduced management overhead

- Ephemeral and stateless nature, with no direct control over persistent storage or file system configurations

Limitations for Jenkins in Serverless Environments

Not all applications are suitable for serverless environments. **Jenkins**, for example, is fundamentally designed as a stateful application, relying on persistent storage, file permissions, and stateful interactions.

Attempting to deploy Jenkins in a serverless platform like Cloud Run or AWS Fargate typically leads to challenges, including

- File permission errors

- Persistent storage and file system conflicts

- Initialization problems due to the ephemeral nature of containers

This incompatibility arises because serverless platforms provide ephemeral containers, restricted and often read-only file systems, and no direct control over persistent volumes or mounts.

Recommended Alternatives for Jenkins

The most suitable environment for running Jenkins remains virtual machines (VMs), as demonstrated in earlier chapters. VMs provide full control over the file system and user permissions and allow the use of persistent disks, which Jenkins requires due to its frequent and intensive use of storage.

An alternative for specific Jenkins tasks, such as running Extract-Transform-Load (ETL) jobs, is to leverage GCP's **Cloud Functions**, which are serverless and well-suited for running scheduled Python scripts. Combined with GCP's **Cloud Scheduler**, a cron-based scheduling service, this approach provides functionality similar to Jenkins' scheduling features without infrastructure overhead.

With these concepts in mind, let's proceed to deploy suitable applications on a serverless infrastructure using Google Cloud Run.

Getting Started with Google Cloud Run

In this section, we will deploy the Flask and Streamlit applications using Google Cloud Run. Follow the detailed steps below to get your environment prepared and authenticated.

Preparing Your Environment

1. **Open Google Cloud Shell Editor**

    - In your Google Cloud Platform (GCP) console, open the Google Cloud Shell Editor, a VS Code-like environment. (Refer to Figure 8-8 for guidance.)

2. **Upload application folders**

- Upload your flask and streamlit application directories separately. Your working directory should resemble the structure from Code-block 9-1.

Code-block 9-1. *Folder structure*

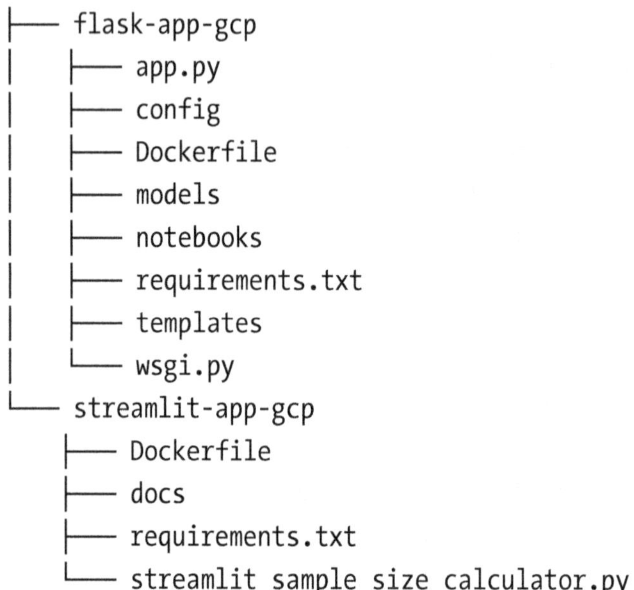

```
├── flask-app-gcp
│   ├── app.py
│   ├── config
│   ├── Dockerfile
│   ├── models
│   ├── notebooks
│   ├── requirements.txt
│   ├── templates
│   └── wsgi.py
└── streamlit-app-gcp
    ├── Dockerfile
    ├── docs
    ├── requirements.txt
    └── streamlit_sample_size_calculator.py
```

Note Internal Application Ports   Earlier in the book we ran **Streamlit on port 8501** and **Flask on port 8502**. In Google Cloud Run, however, every container must listen on **port 8080**. Cloud Run forwards external traffic to that internal port automatically.

To meet this requirement, the repository already includes updated files:

- flask-app-gcp/app.py

- flask-app-gcp/Dockerfile

- streamlit-app-gcp/Dockerfile

Each of these now binds the application to **8080** so the services run correctly in a serverless Cloud Run environment.

Setting Up Environment Variables and Authentication

Run the following commands in your Cloud Shell terminal to set necessary environment variables and authenticate with Google Cloud; see Code-block 9-2.

Code-block 9-2. *Initial commands*

```
export GOOGLE_CLOUD_PROJECT=<YOUR_PROJECT>
export LOCATION=<YOUR_REGION>
gcloud auth login
```

- Replace <YOUR_PROJECT> with your actual Google Cloud project ID (e.g., lb-cloud-project).

- Replace <YOUR_REGION> with your preferred region (e.g., us-west2).

Upon running gcloud auth login, you will be prompted to authenticate via a browser:

1. Click the provided link in the terminal.

2. Follow the authentication steps: click **Open**, **Continue**, and **Allow**.

3. Copy the generated code and paste it back into your terminal.

4. Press **Enter** to complete the authentication.

Your environment is now ready, and you are authenticated with the Google Cloud SDK; see Figure 9-1.

Figure 9-1. *Terminal on GCP shell editor*

Deploying Cloud Run Services Individually

With your environment set up and authenticated, you're ready to deploy your Flask and Streamlit applications separately using Google Cloud Run. Follow the detailed steps below for each deployment.

Deploying the Flask Application

Execute the following commands in your Cloud Shell terminal to build and deploy your Flask application; see Code-block 9-3.

Code-block 9-3. *Deploy Flask application with Cloud Run*

```
gcloud builds submit flask-app-gcp/ --tag gcr.io/$GOOGLE_CLOUD_PROJECT/
flask_app --region=$LOCATION

gcloud run deploy flask-app \
  --image gcr.io/$GOOGLE_CLOUD_PROJECT/flask_app \
  --region=$LOCATION --allow-unauthenticated
```

Understanding the Deployment Commands

The first command, **gcloud builds submit**, takes the contents of the flask folder that you've uploaded and builds a container image. This image is then stored in the Google Cloud Platform's Artifact Registry (https://console.cloud.google.com/artifacts). Inside this registry, you will find a folder named gcr.io, within which your Flask application artifact (flask_app) is saved as a container image. By default, this command tags the built image as "latest," but previous artifacts remain accessible in the registry, allowing you to revert to an earlier version of the app if needed. Additionally, each build submitted to the Artifact Registry is automatically logged and saved in the Google Cloud Build history (https://console.cloud.google.com/cloud-build/builds). This comprehensive logging allows easy access to past builds for troubleshooting, reuse, or version control.

The second command, **gcloud run deploy**, takes the built image artifact you've created and deploys it to Google Cloud Run, a fully managed, serverless environment provided by Google Cloud Platform. Once deployed, your Flask application becomes accessible through Google Cloud Run's service dashboard (https://console.cloud.google.com/run). Here, you can click on your deployed flask-app service to view its unique URL, for example, `https://flask-app-111395551744.us-west2.run.app`, and access additional features. These features include detailed metrics for monitoring request counts, latency, CPU usage, logs, and other essential performance indicators. See Figure 9-2.

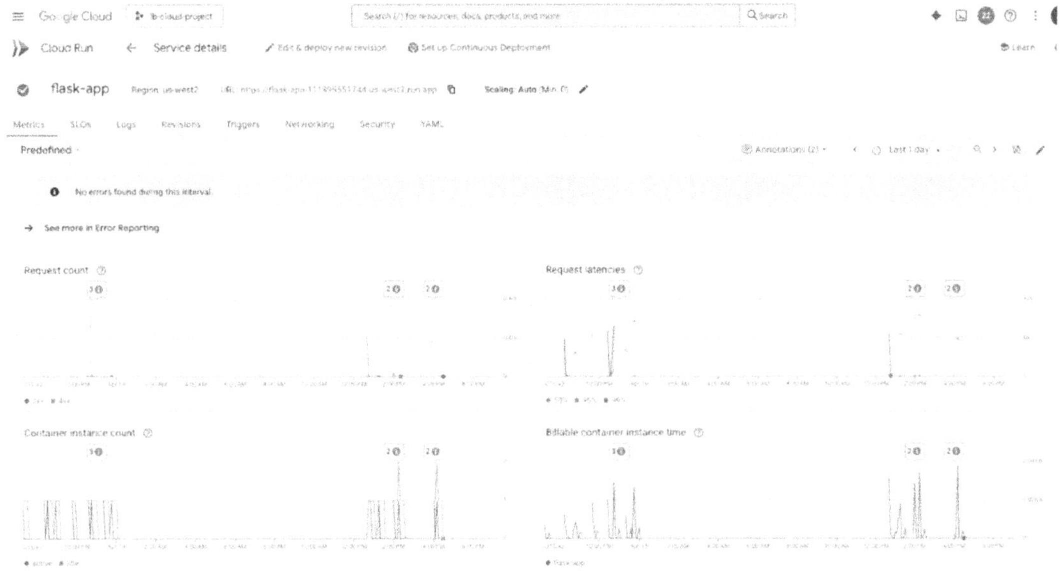

Figure 9-2. *Flask-app on Cloud Run*

Configuring Autoscaling

One important configuration option available for your deployed Cloud Run services is setting the minimum autoscaling instances. By default, Cloud Run sets this to 0, meaning no instances are running until a request comes in. Setting the minimum number of instances to at least 1 ensures that your application always has at least one instance running, significantly reducing initial response times. This setting is particularly important in latency-sensitive environments, such as processing payment transactions. See Figure 9-3 for visual guidance on setting this configuration.

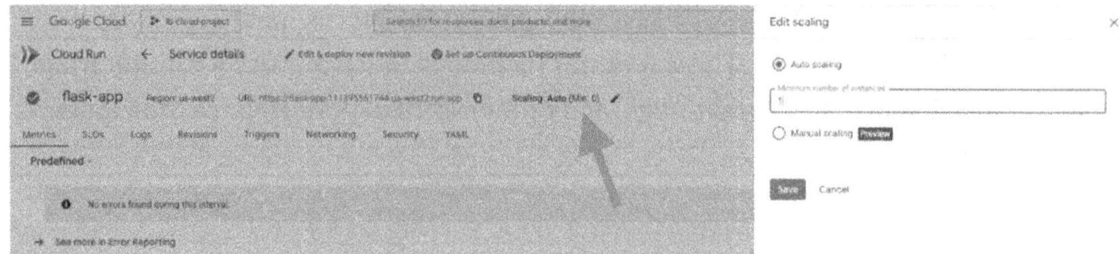

Figure 9-3. *Autoscaling minimum number of instances*

Deploying the Streamlit Application

Now proceed to deploy your Streamlit application. Use the following commands in your Cloud Shell terminal; see Code-block 9-4.

Code-block 9-4. Deploy Streamlit application with Cloud Run

```
gcloud builds submit streamlit-app-gcp/ --tag gcr.io/$GOOGLE_CLOUD_PROJECT/
streamlit_app --region=$LOCATION

gcloud run deploy streamlit-app \
   --image gcr.io/$GOOGLE_CLOUD_PROJECT/streamlit_app \
   --region=$LOCATION --allow-unauthenticated
```

These commands mirror the Flask deployment process. The first command creates a container image from your uploaded `streamlit` directory and stores it similarly in the Artifact Registry. The second command deploys this image to Google Cloud Run, enabling your Streamlit application to become publicly accessible via its unique URL, which you can find in your Cloud Run dashboard.

At this stage, both your Flask and Streamlit applications are successfully deployed as independent serverless Cloud Run services.

Create Serverless Network Endpoint Groups (NEGs)

With your Flask and Streamlit applications successfully deployed on Cloud Run, the next step is to create serverless Network Endpoint Groups (NEGs). These NEGs enable your applications to connect seamlessly to a load balancer. In the previous chapter (Chapter 8), you utilized instance groups for virtual machine deployments. For serverless deployments with Cloud Run, NEGs serve as the analogous connection point required for integration with Google Cloud's load balancing services.

Step-by-Step Instructions for Creating NEGs

You will create two separate NEGs: one for Flask (`flask-neg`) and one for Streamlit (`streamlit-neg`). Execute the following commands from Code-block 9-5 in your Cloud Shell terminal.

Code-block 9-5. Creating Network Endpoint Group (NEG)

```
gcloud compute network-endpoint-groups create flask-neg \
  --region=$LOCATION \
  --network-endpoint-type=serverless \
  --cloud-run-service=flask-app

gcloud compute network-endpoint-groups create streamlit-neg \
  --region=$LOCATION \
  --network-endpoint-type=serverless \
  --cloud-run-service=streamlit-app
```

Explanation of Commands

- **`gcloud compute network-endpoint-groups create flask-neg`**

  - Creates a serverless Network Endpoint Group named `flask-neg`

  - Specifies the region as your previously defined location (`$LOCATION`)

  - Defines the endpoint type as `serverless`, indicating it will connect to serverless services rather than VMs

  - Connects this NEG explicitly to your previously deployed Cloud Run service named `flask-app`

- **`gcloud compute network-endpoint-groups create streamlit-neg`**

  - Creates a serverless Network Endpoint Group named `streamlit-neg`

  - Specifies the region using the same previously defined location (`$LOCATION`)

- Sets the endpoint type to `serverless`

- Connects this NEG to the previously deployed Cloud Run service named `streamlit-app`

These NEGs serve as the vital link enabling the load balancer to direct incoming traffic to your respective Flask and Streamlit serverless applications, enhancing your deployment's scalability and reliability.

SSL Certificates in Google Cloud Platform (GCP)

The next essential step in securing your Cloud Run applications is uploading SSL certificates to Google Cloud Platform. This process has previously been detailed in Chapter 8, section "Set Up a Global Load Balancer with SSL, HTTPS, and Subdomain Support," and the steps remain consistent for serverless deployments. Ensure your configuration matches the following folder structure; see Code-block 9-6.

Code-block 9-6. Config folder structure

```
├── config
│   ├── lb-webserver_pro_combined.crt
│   └── private
```

Make sure your SSL certificates (`lb-webserver_pro_combined.crt`) and private key (`private`) files are correctly placed within the `config` folder. Refer back to Chapter 8, section "Set Up a Global Load Balancer with SSL, HTTPS, and Subdomain Support," if additional guidance or a detailed review of the upload process is needed.

Creating the Load Balancer

Now it's time to create the load balancer, which will include two backend services, one for each of our deployed serverless applications: Flask and Streamlit. The type of load balancer and the frontend configuration will closely resemble what we previously set up in Chapter 8, section "Set Up a Global Load Balancer with SSL, HTTPS, and Subdomain Support."

Step-by-Step Instructions

1. **Initiate load balancer creation**

   - In the Google Cloud Console, navigate to **Network Services ➤ Load Balancing**.

   - Click CREATE LOAD BALANCER.

   Configure using these default settings, suitable for our global deployment:

   - **Type of load balancer:** Select **Application Load Balancer (HTTP/HTTPS)**; click NEXT.

   - **Public facing or internal:** Choose **Public facing (external)**; click NEXT.

   - **Global or single region deployment:** Choose **Best for global workloads**; click NEXT.

   - **Load balancer generation:** Select **Global external Application Load Balancer**; click NEXT.

   - Then click CONFIGURE. See Figure 9-4.

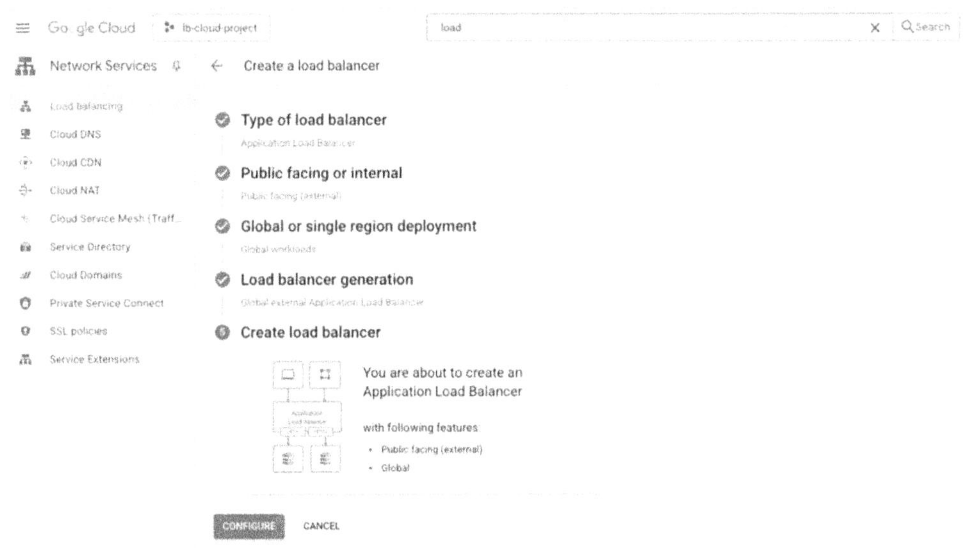

Figure 9-4. *Configure load balancer*

Give the load balancer a name, for example, `lb-https`.

2. **Frontend configuration** *(similar to Chapter 8, section "Set Up a Global Load Balancer with SSL, HTTPS, and Subdomain Support," Figure 8-10)*

   Configure the frontend as follows:

   - **Name:** Enter a descriptive name, for example, `lb-https-frontend`.

   - **Protocol:** Select **HTTPS (includes HTTP/2 and HTTP/3)**.

   - **IP version:** Select **IPv4**.

   - **IP address:** Click the drop-down, select **CREATE IP ADDRESS**, name it clearly (e.g., `ip-cloudrun`), and click **RESERVE**.

   - **Port:** Enter 443.

   - **Certificate:** Select your previously uploaded SSL certificate (`lb-webserver-cert`).

   - After completing these settings, click **DONE**.

3. **Backend configuration** *(specific for serverless Cloud Run services)*

   Now we need to create two distinct backend services: one for Flask and one for Streamlit. Unlike in the previous chapter (Chapter 8), here we'll use serverless Network Endpoint Groups (NEGs).

   Create Backend Service for Flask

   - Click **Backend services & backend buckets**, then select **CREATE A BACKEND SERVICE**:

     - **Name:** `lb-https-backend-flask`

     - **Backend type:** Select **Serverless network endpoint group**.

     - **Backends:** Select `flask-neg`, then click **DONE**.

     - *(Do NOT add another backend here; we'll create a separate one for Streamlit.)*

     - **Cloud CDN:** Uncheck this box because CDN is incompatible with IAP, a service we'll use later for authentication.

- Scroll to **Security**: Leave default values.

  - **Policy name:** Google Cloud may auto-generate a policy name exceeding 63 characters (which is invalid). Shorten this auto-generated policy name if necessary to under 63 characters.

- Click **CREATE** to save the backend service.

Create Backend Service for Streamlit

- Again click **CREATE A BACKEND SERVICE:**

  - **Name:** `lb-https-backend-streamlit`

  - **Backend type:** Select **Serverless network endpoint group**.

  - **Backends:** Select `streamlit-neg`, then click **DONE**.

  - **Cloud CDN:** Uncheck this box for the same reason (incompatibility with IAP).

- Click **CREATE** to finalize the second backend service.

Refer to Figure 9-5 for visual confirmation.

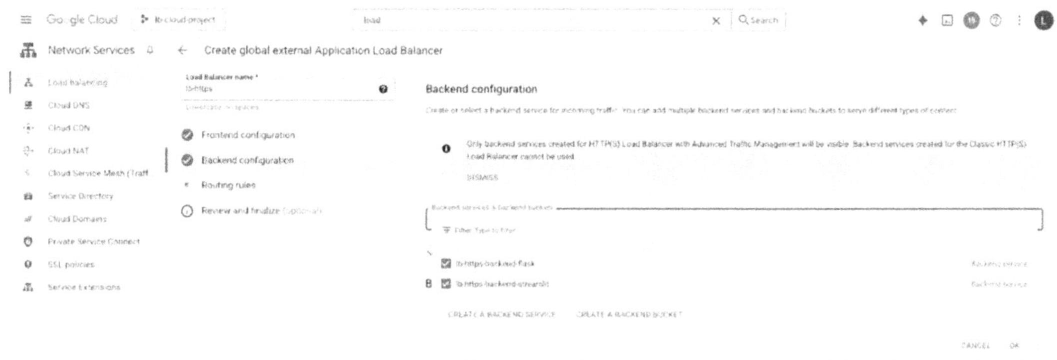

Figure 9-5. *Two separate backends created*

4. **Adjust routing rules** *(to direct requests properly)*

   Before finalizing the load balancer, you must tweak the routing
   rules to ensure traffic is correctly directed to the respective
   backend services. Adjust these rules following the guidance
   provided in Figure 9-6.

   - **Rule 1**

     - First rule is the default; just choose a backend, either flask or
       streamlit, that will be directed there in case it doesn't match
       any specific rules that will be defined by you in rules 2 and 3.

   - **Rule 2**

     - **Host 2:** streamlit.lb-webserver.pro

     - **Path 2:** /*

     - **Backend 2:** lb-https-backend-streamlit

   - **Rule 3**

     - **Host 3:** flask.lb-webserver.pro

     - **Path 3:** /*

     - ***Backend 3:*** lb-https-backend-flask

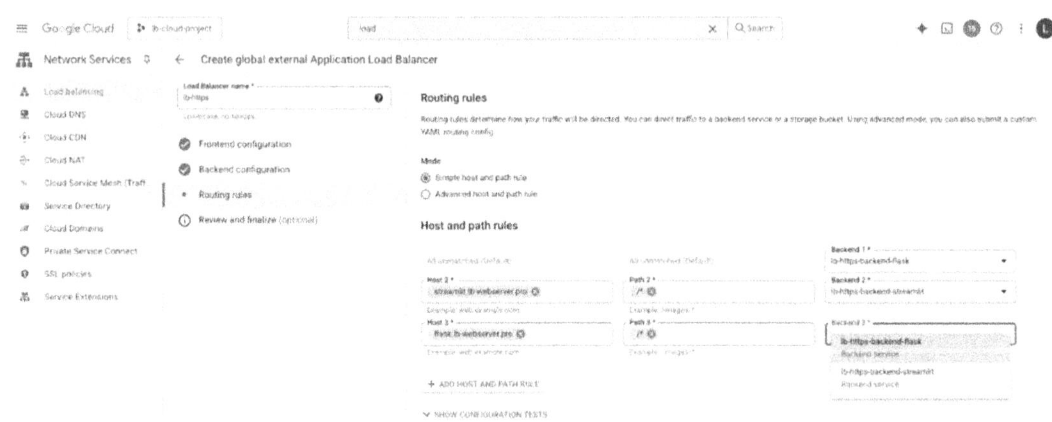

Figure 9-6. *Adjusting routing rules*

5. **Finalize and create the load balancer**

- Review your configuration to confirm accuracy.

- Click **CREATE** at the bottom-left corner.

- Wait for the load balancer provisioning to complete.

6. **Retrieve Load Balancer IP** *(needed for DNS configuration)*

Once the load balancer is successfully created:

- Click on the Load Balancer name.

- Navigate to the **Frontend** section and find the assigned **IP address**.

(In our example, the IP is 34.8.41.246, as shown in Figure 9-7.)

Figure 9-7. *Get the IP from Load Balancer Frontend*

Note   Google's HTTPS Load Balancer includes automatic Layer 3 and 4 DDoS protection as part of its globally distributed infrastructure, requiring no further configuration.

Now you have successfully configured your load balancer with two distinct backend services for your Flask and Streamlit applications running on Google Cloud Run. You're ready to continue to the next steps for securing and configuring user authentication.

Updating DNS Records on Namecheap

With the load balancer now fully operational and the external IP address available, you need to update your DNS records on Namecheap. This step ensures that traffic is correctly routed to your Flask and Streamlit applications through the designated subdomains:

- **flask.lb-webserver.pro**

- **streamlit.lb-webserver.pro**

To perform this update:

1. Log in to your account at **namecheap.com**.

2. Navigate to your DNS management settings for the domain **lb-webserver.pro**.

3. Replace the current IP addresses with the new Load Balancer IP (e.g., 34.8.41.246) for the Flask and Streamlit subdomains.

Note   Do not update the IP address for unrelated services (e.g., Jenkins), since they aren't deployed with Cloud Run at this stage.

Refer to Figure 9-8 for a visual guide on adjusting DNS settings.

	Type	Host	Value
	A Record	@	34.8.41.246
	A Record	flask	34.8.41.246
	A Record	jenkins	18.228.131.1
	A Record	streamlit	34.8.41.246

Figure 9-8. *Adjust DNS on Namecheap*

Restricting Direct Access to Cloud Run URLs

To enhance security and ensure your applications are accessed only through the load balancer, you need to restrict direct access to the Cloud Run service URLs. Google Cloud provides an ingress setting (`internal-and-cloud-load-balancing`) that ensures the service can only be accessed internally within Google Cloud or through Cloud Load Balancers.

You can apply these restrictions by running the commands from Code-block 9-7 in your terminal or Cloud Shell.

Code-block 9-7. Block direct access to Cloud Run apps

```
gcloud run services update flask-app \
  --region=us-west2 \
  --ingress internal-and-cloud-load-balancing

gcloud run services update streamlit-app \
  --region=us-west2 \
  --ingress internal-and-cloud-load-balancing
```

Once these settings are applied, you'll notice the URLs for your Cloud Run services appear faded and nonclickable in the Google Cloud Console interface. This indicates that direct public access has been successfully disabled.

Refer to Figure 9-9 for an example of the Cloud Run URL becoming faded after applying these settings.

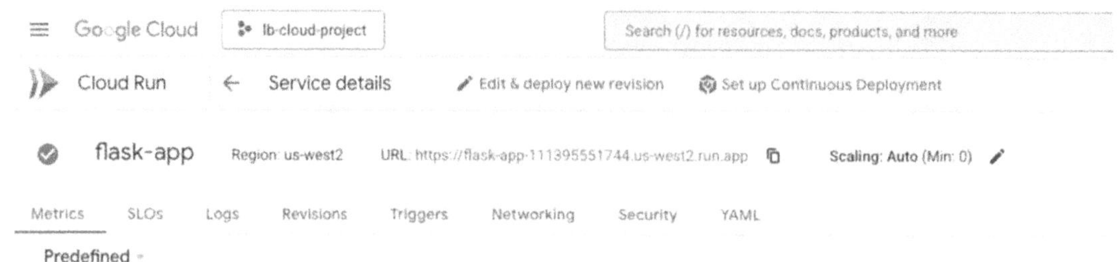

Figure 9-9. *Cloud Run URL faded out*

Configuring Authenticated Access

To ensure secure, authenticated access to your deployed applications through the load balancer, you can use Identity-Aware Proxy (IAP) and, optionally, Cloud Armor for IP-based restrictions.

Identity-Aware Proxy (IAP)

Although your applications are now accessed exclusively via the load balancer, they remain publicly accessible. Implementing Identity-Aware Proxy (IAP) ensures only authenticated users can access your applications.

Steps to Enable IAP

1. In the Google Cloud Console, navigate to **Security ➤ Identity-Aware Proxy**.

2. You will see your backend services listed here. If an error appears indicating the need for configuration, resolve it as prompted; see Figure 9-10.

Figure 9-10. *Identity-Aware Proxy page*

3. Enable IAP by clicking the toggle under the ***IAP*** column next to your backend services. Agree to the configuration terms and click ***TURN ON***; see Figure 9-11.

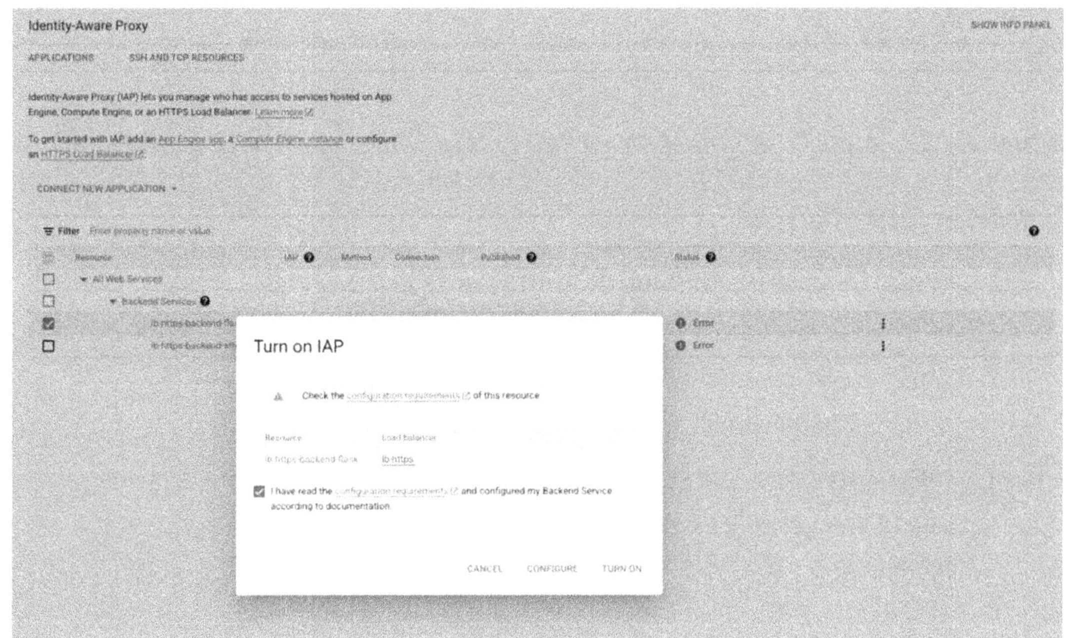

Figure 9-11. *Turn on IAP*

4. After enabling IAP, wait a few minutes and then access your
    application domain (e.g., `flask.lb-webserver.pro`). You will
    now see Google's sign-in page prompting for authentication; see
    Figure 9-12.

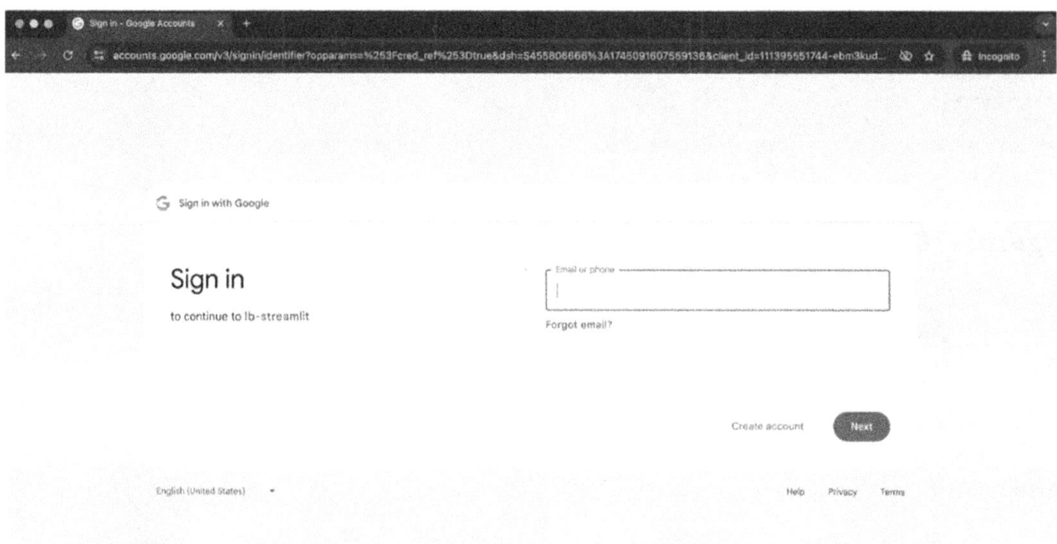

Figure 9-12. *Access flask.lb-webserver.pro and see the sign-in page from Google.*

Managing User Access

1. Return to **Security ➤ Identity-Aware Proxy** in the Google Cloud
    Console.

2. Select the backend service for which you want to manage access.
    A right-side panel will appear.

3. Click **ADD PRINCIPAL**, enter user emails, and assign them the
    role **IAP-secured Web App User** to grant authenticated access;
    see Figure 9-13.

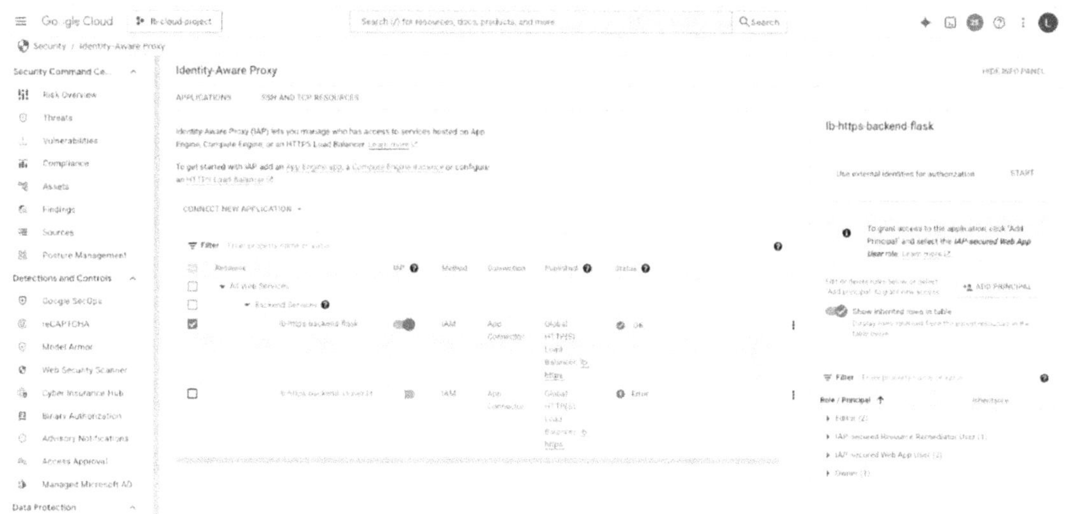

Figure 9-13. *Grant access to users to pass the IAP sign-in page*

You can repeat this process for any other deployed applications, including your Streamlit application.

Choosing the Right Protection Method

- **Identity-Aware Proxy (IAP)** is ideal for browser-based access to your user interfaces—for example, `https://flask.webserver.pro` and `https://streamlit.webserver.pro`.

  It presents a Google sign-in page and restricts entry to the users you whitelist.

- For machine-to-machine calls—such as the Flask prediction API at `https://flask.webserver.pro/predict`—IAP can be cumbersome. In that case, it's simpler to secure the endpoint with **Cloud Armor**, which lets you whitelist specific IP addresses or ranges.

The next section walks you through configuring Cloud Armor for that purpose.

Cloud Armor for IP-Based Security

Alternatively, you can restrict access to your applications by whitelisting specific IP addresses using Cloud Armor policies.

Steps to Configure Cloud Armor

1. In the Google Cloud Console, navigate to **Network Security ➤ Cloud Armor policies** and click **Create policy**; see Figure 9-14.

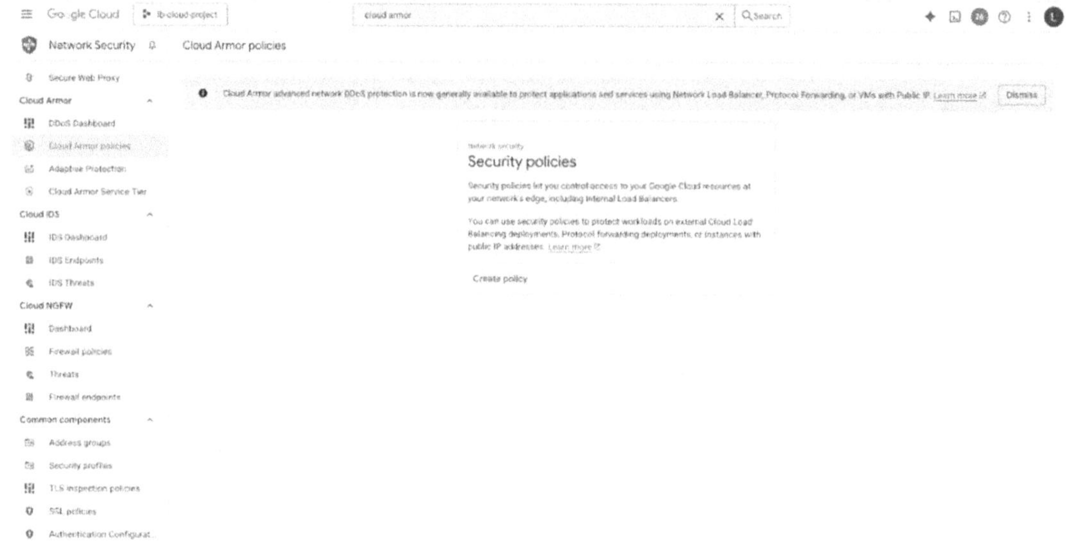

Figure 9-14. *Cloud Armor policies*

2. Configure the policy.

   - **Name:** Use a descriptive name, such as restrict-by-ip.

   - **Policy type:** Select **Backend security policy**.

   - **Scope:** Choose **Global**.

   - **Policy layer:** Select **Application**.

   - **Default rule action:** Select **Deny**, with a response code of **403 (Forbidden)**. Click **Next step**.

3. Add IP whitelist rule.

   - **Description:** Clearly describe the rule, for example, whitelist specific IPs.

   - **Mode:** Choose **Basic mode**.

- **Match condition:** Enter your specific IP or IP range (e.g., `87.154.215.35/32`).

- **Action:** Select **Allow**.

- **Priority:** Set priority (lower number means higher priority, e.g., 1000).

- Click **Next step**.

4. Apply policy to backend services (click `Add target`).

    - **Type:** Select **Backend service**.

    - Add each backend service (e.g., `lb-https-backend-flask` and `lb-https-backend-streamlit`).

    - Click **Next step**.

5. Review settings and click **Create policy** to finalize.

Note   If you encounter a **quota exceeded** error (see Figure 9-15), this means your account's Cloud Armor quota is set to zero by default. New personal accounts typically need to request a quota increase from Google Cloud support.

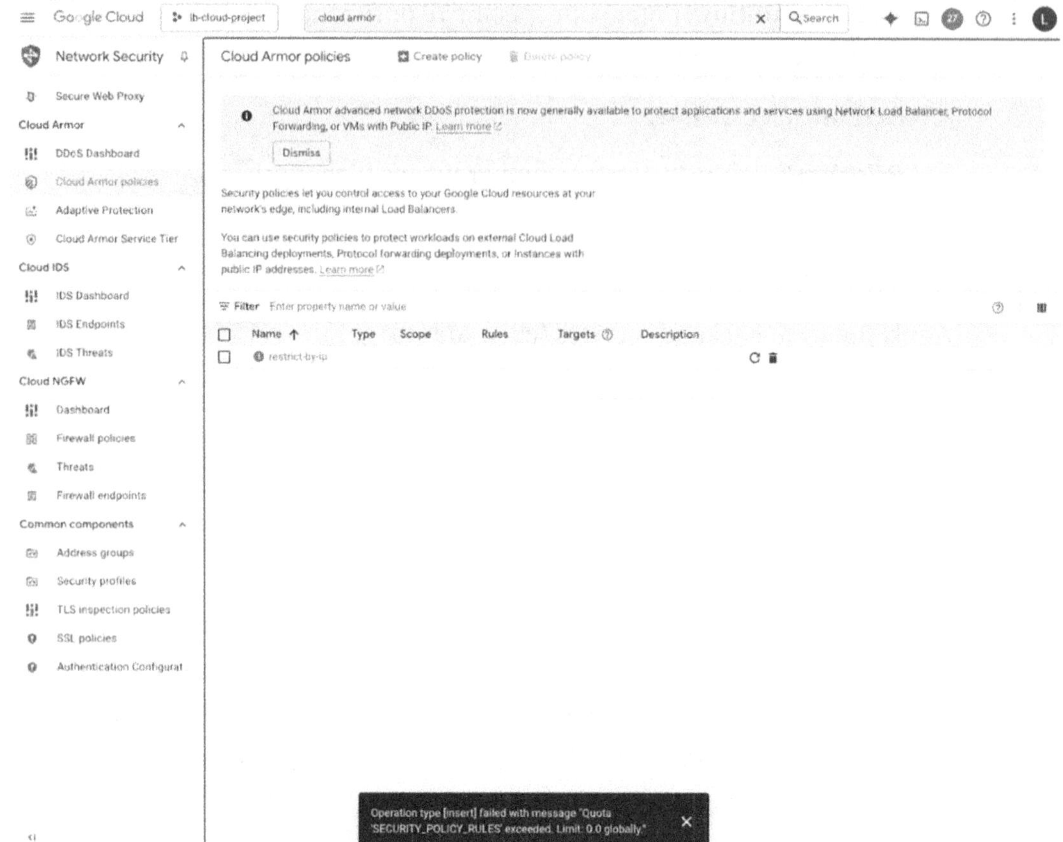

Figure 9-15. *Quota exceeded limit*

How to Request a Quota Increase from Google Cloud

Sometimes Cloud Armor setup fails with **Quota exceeded** errors because new projects start with **zero** allowed security-policy rules. Before you contact Google, confirm that this is indeed the problem.

Step 1: Verify the current quota

1.   Navigate to **IAM & Admin ➤ Quotas & System Limits**.

2.   In the **Search** box, type `security_policy_rules`.

3.   A long list appears, one entry per region plus one "global" entry.

4.   Sort by **Region** and locate the row that has **no specific region** (Service = *Compute Engine API*, Type = *Quota*). If the value is **0**, you have found the culprit; see Figure 9-16.

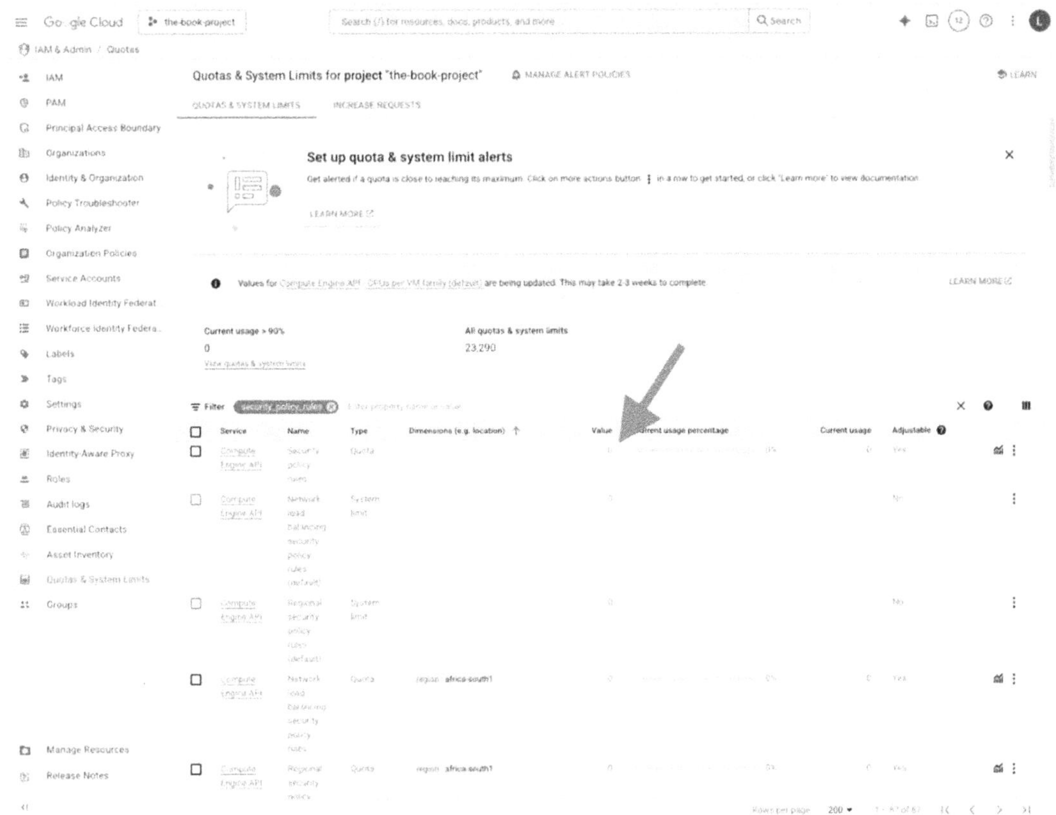

Figure 9-16. *Security policy rules value set to 0*

Step 2: Contact Google Cloud Sales

Google raises this quota manually. Go to **https://cloud.google.com/contact** and choose one of two options:

Contact method	How it works
Chat online with us	Live text chat during the business hours shown on the page.
Request a call back	Fill in a short form; you can opt for email instead of a phone call.

When the chat window opens (see Figure 9-17), pick **Live chat with sales** and describe the request, for example:

*"I need to raise Security policy rules quota to **100** for the Compute Engine API."*

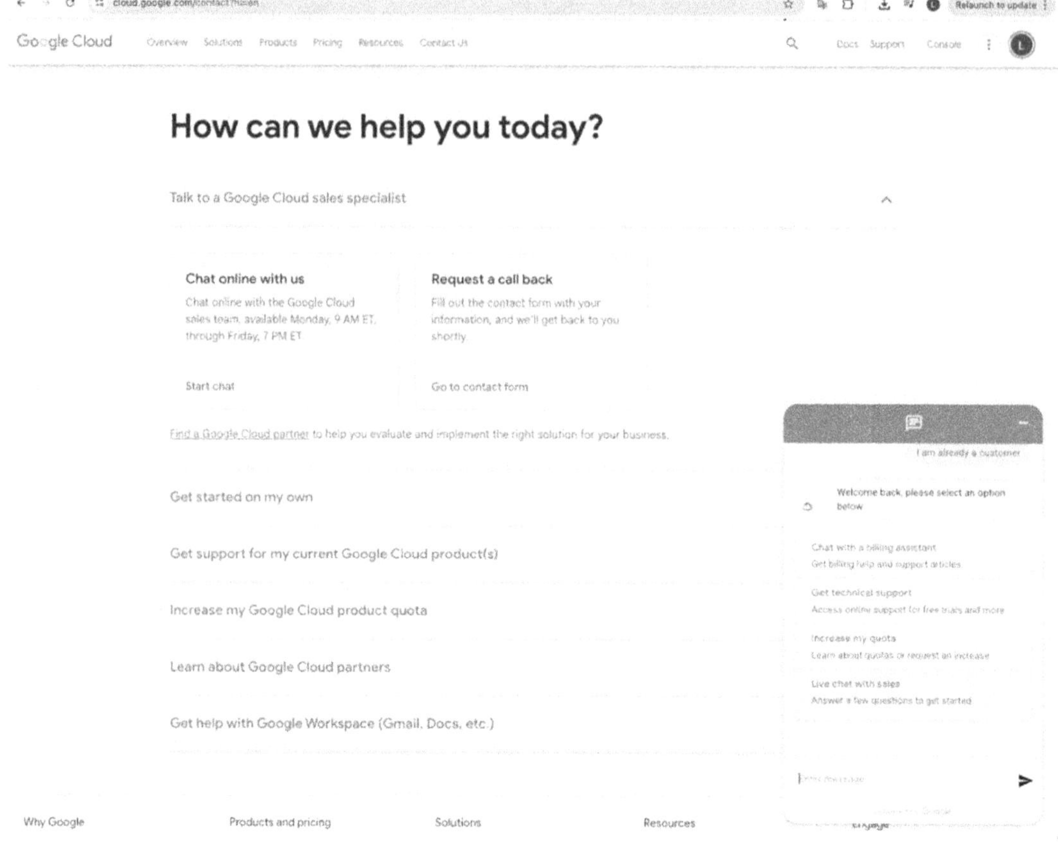

Figure 9-17. *Reaching out to Google Cloud Sales Team*

Expect a few profile questions: company name and role if you're representing a business, or confirmation that this is a personal project.

- If you name a company, Google Sales representative routes you to the account's sales representative that will reach out to you within two to three business days.

- For personal projects, Google Sales representative may direct you to a local **Google Cloud Partner** (*https://cloud.google.com/find-a-partner*) who will submit the quota request on your behalf.

Step 3: Confirm the increase

To double-check whether your quota has indeed increased, return to **IAM & Admin** ➤ **Quotas & System Limits**, search again for **security_policy_rules**, and verify the value has been updated; see Figure 9-18. You can now create Cloud Armor rules without hitting the limit.

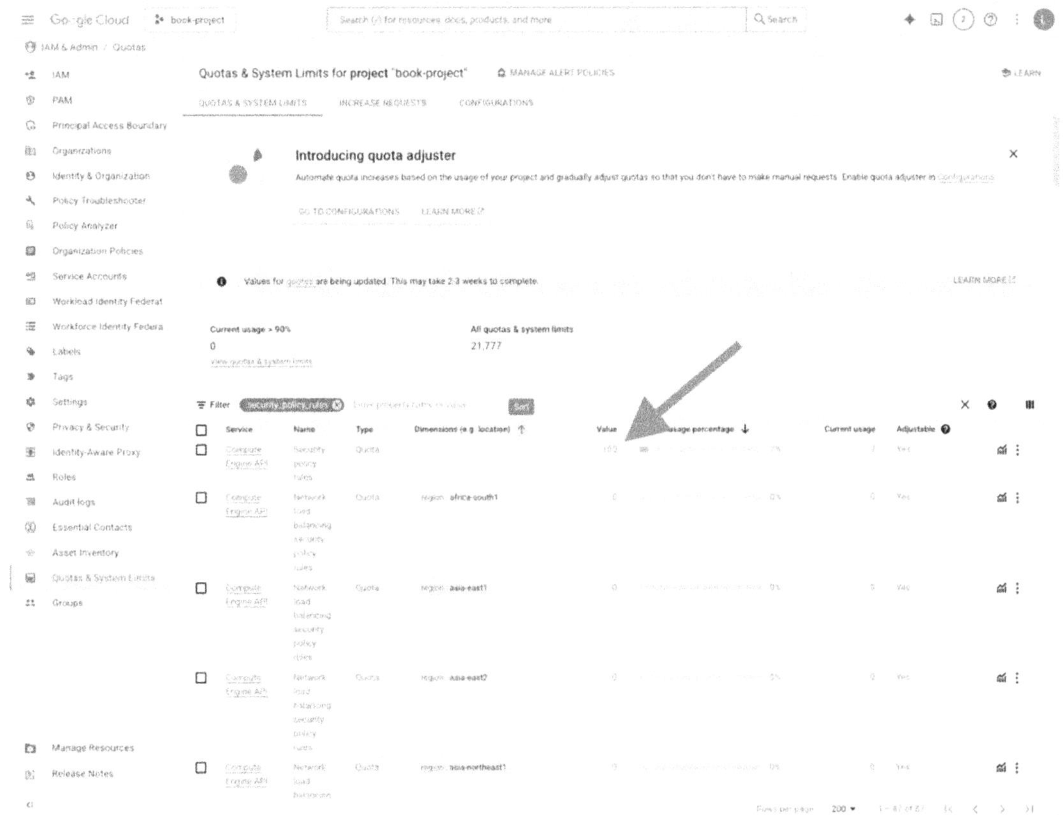

Figure 9-18. *Security policy rules value set to 100*

Summary

In this chapter, you deployed Flask and Streamlit applications to a serverless infrastructure using Google Cloud Run and secured them with HTTPS and authentication layers. Specifically, you

- Reviewed key concepts, what is serverless, differences from VM-based deployment, and why Jenkins is not suitable for Cloud Run

  - Google Cloud Serverless documentation: `https://cloud.google.com/serverless`

- Deployed Flask and Streamlit as individual Cloud Run services, both configured to run on port 8080

  - Google Artifact Registry documentation: `https://console.cloud.google.com/artifacts`

- Built Docker images using `gcloud builds submit` and deployed them with `gcloud run deploy`

  - Google Cloud Build documentation: `https://console.cloud.google.com/cloud-build/builds`

- Created **serverless Network Endpoint Groups (NEGs)** to route traffic from the load balancer to your Cloud Run services

  - Serverless NEGs documentation: `https://cloud.google.com/load-balancing/docs/negs/serverless-neg-concepts`

- Uploaded and configured **SSL certificates** to secure your services with HTTPS

  - Google Cloud SSL Certs documentation: `https://cloud.google.com/load-balancing/docs/ssl-certificates`

- Created a global HTTPS **Load Balancer** with two backends: Flask and Streamlit, each connected to its NEG

- Configured **routing rules** based on hostname (flask.lb-webserver.pro and streamlit.lb-webserver.pro)

- Updated **DNS records on Namecheap** to point subdomains to your Load Balancer IP

- Restricted direct Cloud Run access using `--ingress internal-and-cloud-load-balancing`

- Enabled **Google Identity-Aware Proxy (IAP)** for email-based authenticated access

  - Identity-Aware Proxy documentation: `https://cloud.google.com/iap/docs`

- Configured **Cloud Armor policies** to restrict access to specific IPs for API endpoints

  - Cloud Armor documentation: `https://cloud.google.com/armor/docs`

In the next chapter, you'll apply similar serverless deployment strategies using AWS. You'll create IAM users and access keys, configure the AWS CLI, create ECR repositories, and push Docker images. You'll prepare the environment for ECS + Fargate, deploy your containers, enable HTTPS using your own certificates, and scale the applications with minimal infrastructure management.

## CHAPTER 10

Serverless Deployment with AWS

In Chapters 1–6, you built an end-to-end production environment on AWS entirely from scratch, networks, subnets, security groups, EC2 instances, an Application Load Balancer (ALB), and the whole stack.

Now you will achieve the same functional outcome **without** provisioning or managing any servers yourself. We are moving to a *serverless* architecture, letting AWS run and scale the undifferentiated heavy-lifting for us. Some of the benefits from using serverless infrastructure are faster time to market, no patching, sizing, or Operational System (OS) maintenance.

- Pay-as-you-go, costs follow actual usage.

- Built-in elasticity, autoscales to zero and up again.

Why still learn the basics? Because a good engineer understands how each part of a system works, making it much easier to troubleshoot a managed service when you've once built and understood its underlying components by hand.

By the end of this chapter, you will have two applications, **Flask** and **Streamlit**, running on AWS ECS Fargate behind an HTTPS load balancer, deployed straight from source with zero server management.

Getting Started

Clone the chapter resources; back in Chapter 2, we've seen how we clone a repository. Table 10-1 shows the changes between the repository for Chapter 10 compared to how it was in Chapters 1–6.

Table 10-1. *What's different in this folder?*

Previously (Chapters 1–6)	Now (Chapter10)	Rationale
jenkins/ directory	*Absent*	Jenkins is stateful and needs persistent storage, unsuitable for a pure serverless demo
docker-compose.yml	*Absent*	Port coordination and network wiring are handled by ECS Fargate and the AWS console
nginx/ reverse proxy	*Absent*	An AWS Application Load Balancer (ALB) terminates TLS and routes traffic for us

Application Layout

The repository now contains only two top-level folders:

- flask-app/

    - **flask-app** – Unchanged code, still listens on port **5000** (the Flask default)

- streamlit-app/

    - **streamlit-app** – Unchanged code, still listens on port **8501** (the Streamlit default)

No port remapping is required; ECS Fargate detects each container's exposed port and configures the service automatically.

With the prerequisites out of the way, the next section walks you through packaging these two containers and deploying them with AWS ECS Fargate.

Create an IAM User and Access Keys

Most of the remaining steps are executed from your terminal with the AWS CLI. For that, you need an **aws_access_key_id** and **aws_secret_access_key**. The secure way to obtain them is to create a dedicated IAM user for this project.

Tip– Least privilege   In production, you would scope the policy set as tightly as possible. For a learning lab, and to avoid friction, we grant full access to the services we'll touch.

Step 1: Create a User Group with the Required Policies

1. Open the AWS Management Console and go to **IAM ➤ Users**.

2. Click **Create user** and enter a descriptive name such as serverless-user.

3. On the *Set permissions* page, choose **Add user to group ➤ Create group**.

4. Attach the following managed policies; see Figure 10-1.

   - AmazonS3FullAccess

   - AmazonEC2FullAccess

   - AmazonEC2ContainerRegistryFullAccess

   - AmazonECS_FullAccess

   - AWSCertificateManagerFullAccess

   - CloudWatchFullAccess

5. Name the group serverless (or similar) and click **Create user group**.

6. Back on the permissions screen, verify the new group is selected, then click ***Create user***.

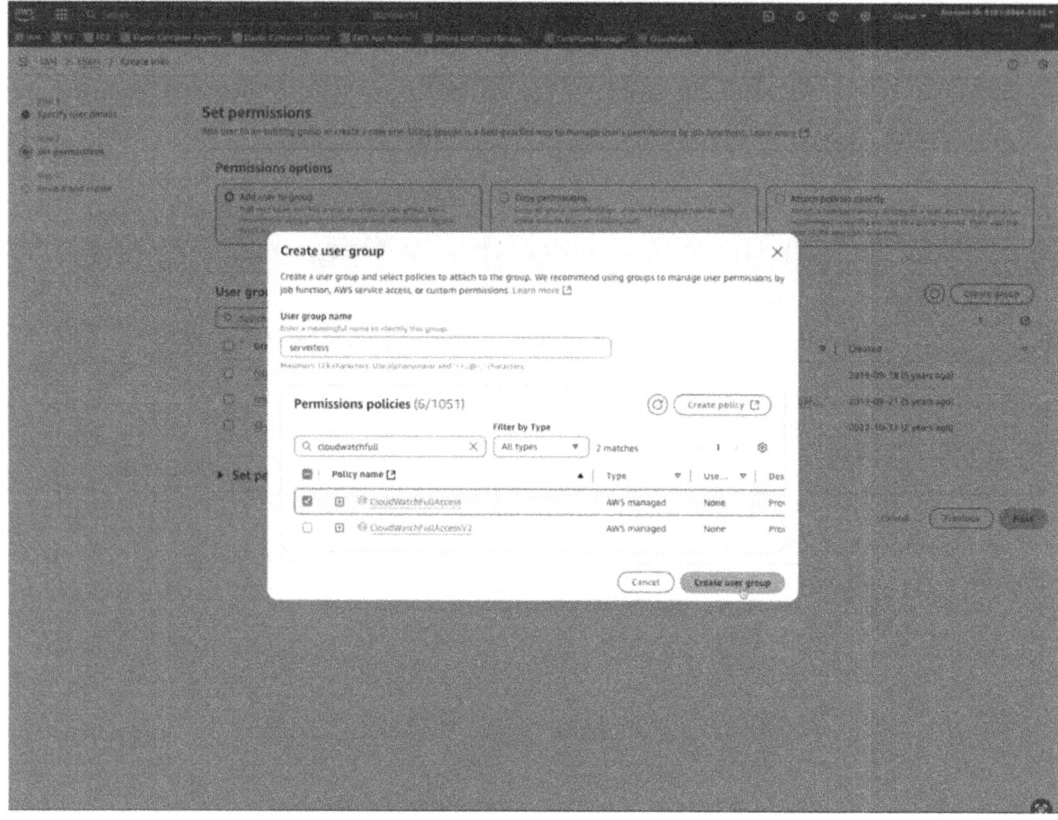

Figure 10-1. *Policies selected*

Step 2: Generate Programmatic Access Keys

1. Select the user you just created.

2. Choose **Security credentials ➤ Create access key**; see Figure 10-2.

3. For *Use case*, select **Command Line Interface (CLI)**, tick the confirmation box, and click **Next**.

4. Optionally add a description (e.g., *local development*), then click **Create access key**.

5. ***Download*** the .csv file or copy the keys to a secure password manager, then click ***Done***.

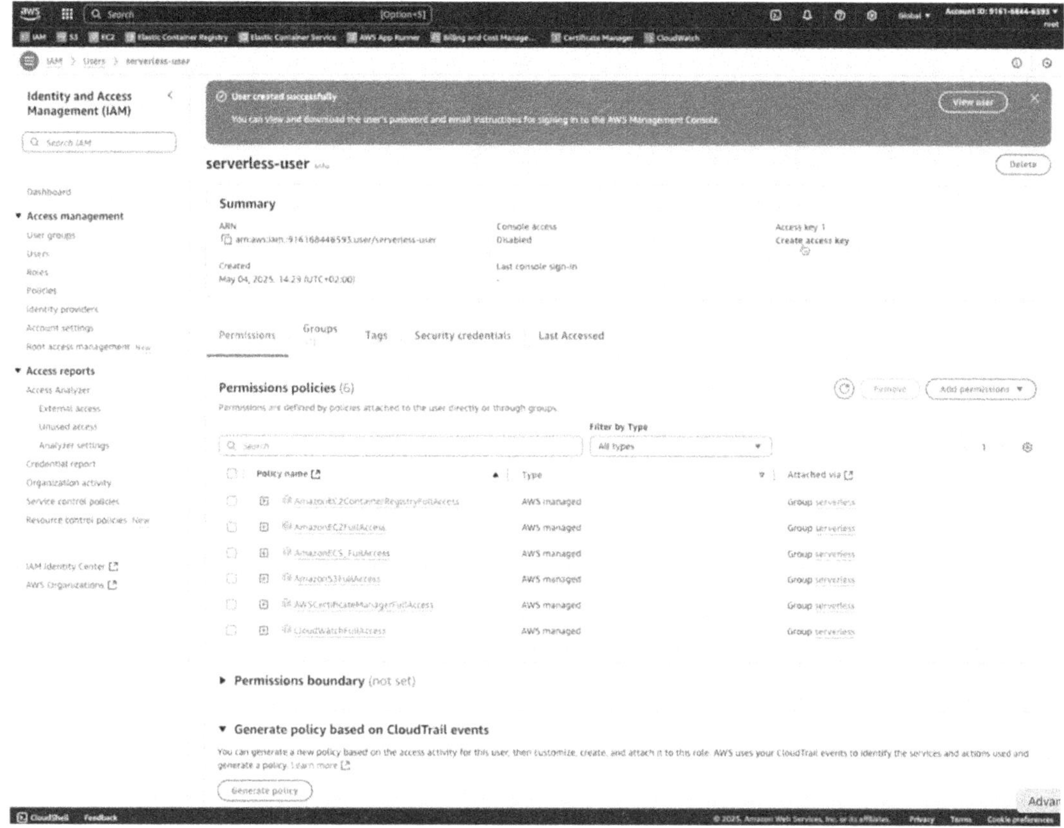

Figure 10-2. *Create access key*

Install and Configure the AWS CLI

The AWS Command Line Interface (CLI) is the workhorse we'll use to push images to Amazon ECR and to automate ECS Fargate deployment. Back to your local VS Code, go to your root folder, mine is `/Users/lucasbraga/`, and check whether you already have it. On your terminal, type and enter `aws --version`; if it returns a version something like **aws-cli/2.27.1**, then you can skip the installation from 10.3.1 and go straight to 10.3.2. *Configure the CLI.* Otherwise if you need to install it, follow Code-block 10-1 for instructions.

Installation

Code-block 10-1. Install AWS CLI

```
# Install on macOS
brew install awscli

# Install on Linux (Ubuntu/Debian/RHEL/etc.)
# Download the installer, unzip it and install it:
curl "https://awscli.amazonaws.com/awscli-exe-linux-x86_64.zip" -o
"awscliv2.zip"
unzip awscliv2.zip
sudo ./aws/install

# Install on Windows
# Download the latest installer for Windows, run it and follow the prompts
# https://awscli.amazonaws.com/AWSCLIV2.msi

# Verify it
aws --version
```

Configure the CLI

Run the interactive wizard and supply the credentials you created in this chapter, section "Create an IAM User and Access Keys"; see Code-block 10-2.

Code-block 10-2. AWS configure

```
aws configure
```

You'll be prompted for four items; see Table 10-2.

Table 10-2. *Configuring AWS CLI*

Prompt	What to enter
AWS Access Key ID	The key from the `.csv` you downloaded
AWS Secret Access Key	The matching secret key
Default region name	For example, `us-east-1` or the region closest to you
Default output format	`json` (or `yaml`, `text`)

The wizard creates two files under **~/.aws/**:

- `credentials` – Stores access keys per profile

- `config` – Stores default output and region per profile

At this point a directory named **~/.aws** exists in your home folder containing two files: `config` and `credentials`. If you prefer to keep everything in a single place, you can copy the key lines from `credentials` into `config` and delete the now-redundant file.

Code-block 10-3 shows a consolidated example:

- **region** – Set this to the AWS region you have been using or the one geographically closest to you.

- The line beginning with # is a comment; it's a handy reminder of which AWS account these keys belong to.

- The stanza in square brackets (`[profile serverless-user]`) defines the **profile name** you will reference when running `aws` commands. Matching the IAM username is a tidy convention.

- Paste the `aws_access_key_id` and `aws_secret_access_key` generated in this chapter, section "Create an IAM User and Access Keys," under that profile.

Code-block 10-3. *"~/.aws/config" file*

```
[default]
region = eu-central-1

# account number 9161123123
[profile serverless-user]
aws_access_key_id = AKIA5AC28B2612
aws_secret_access_key = g+8wm2FZrVhAiuaiUHSjhsiudRH/x
```

Security Reminder

It's good practice to keep these files out of version control. Add ~/.aws/* to your global `.gitignore`.

With the CLI installed and configured, you're ready to package the Docker images and publish them to Amazon ECR in the next section.

Create ECR Repositories and Push Docker Images

AWS Elastic Container Registry (ECR) is a fully managed Docker registry. We'll create one private repository per application and push the images so that AWS ECS Fargate can pull them later. We're doing this still from your VSCode's local terminal.

Export Helper Variables

Make your account-specific values available to the shell (update the placeholders); see Code-block 10-4.

Code-block 10-4. Create environment variables

```
export REGION=us-east-1
export AWS_ACCOUNT_ID=9161123123
```

Create the Repositories

Now let's create **two** ECR repositories, one for the **flask-app** and one for the **streamlit-app**, directly from the command line, as shown in Code-block 10-5. Notice the `--profile serverless-user` flag: it retrieves the credentials associated with the IAM user you created in this chapter, section "Create an IAM User and Access Keys." Those credentials are securely stored in your `~/.aws` folder, so you never hard-code sensitive keys on the command line.

Code-block 10-5. Create two ECR repositories, one for each application

```
# Create an ECR Repository
aws ecr create-repository --repository-name flask-app --profile serverless-user --region $REGION

aws ecr create-repository --repository-name streamlit-app --profile serverless-user --region $REGION
```

> **Why `--profile serverless-user`?**
> This retrieves credentials from `~/.aws/config` rather than exposing them on the command line.

Open **Amazon ECR** in the AWS console (be sure to select the same region) and you should see both repositories listed; see Figure 10-3.

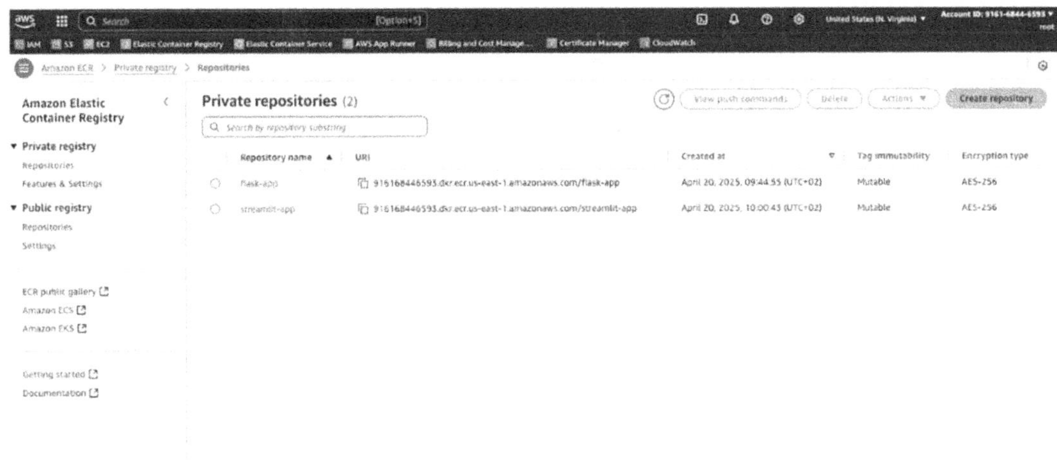

Figure 10-3. *ECR repositories created*

Authenticate Docker to ECR

Docker must authenticate before it can push images. Ensure Docker is installed locally; if not, SSH into a Linux VM that has Docker and repeat these steps (don't forget to re-export $REGION and $AWS_ACCOUNT_ID). See Code-block 10-6.

Code-block 10-6. Authenticate Docker to AWS ECR

```
# Authenticate Docker to AWS ECR
aws ecr get-login-password --region $REGION --profile root-461 | docker
login --username AWS --password-stdin $AWS_ACCOUNT_ID.dkr.ecr.$REGION.
amazonaws.com
```

Build, Tag, and Push the Flask Image

Now that the authentication is done, start creating the Docker images and pushing them to their respective ECR repositories. Start with flask. Move to the flask-app folder and create the Docker image. Refer to Code-block 10-7.

Code-block 10-7. Build the flask Docker image

```
# Build Docker image
cd ~/Documents/GitHub/deploy-secure-ds-apps-book/flask-app/
docker buildx build --platform=linux/amd64 -t flask-app .
```

Tag it with the ECR repo URI and push it to ECR. Refer to Code-block 10-8.

Code-block 10-8. Tag the Docker image and push it to ECR

```
# Tag it
docker tag flask-app:latest $AWS_ACCOUNT_ID.dkr.ecr.$REGION.amazonaws.com/
flask-app:latest

# Push to ECR
docker push $AWS_ACCOUNT_ID.dkr.ecr.$REGION.amazonaws.com/flask-app:latest
```

In the AWS console, click **flask-app** under ECR to confirm the image manifest has arrived; see Figure 10-4.

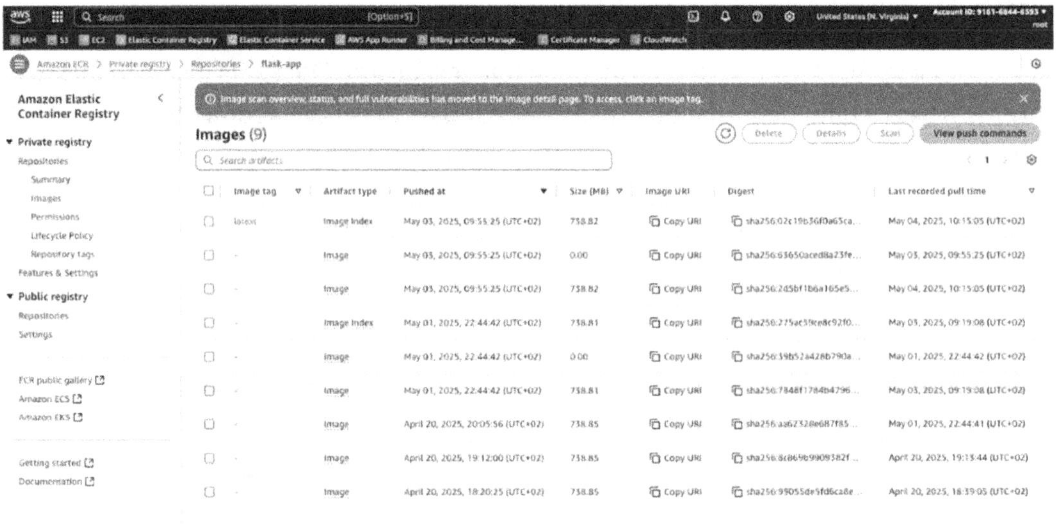

Figure 10-4. *See Docker images pushed to the flask-app ECR repository*

Build, Tag, and Push the Streamlit Image

And do the same for Streamlit; see Code-block 10-9.

Code-block 10-9. Creating the Streamlit Docker image and pushing it to ECR

```
# 3. Build and Push the Docker Image
# Build Docker image
cd ~/Documents/GitHub/deploy-secure-ds-apps-book/streamlit-app/
docker build -t streamlit-app .

# 4. Tag it
docker tag streamlit-app:latest $AWS_ACCOUNT_ID.dkr.ecr.$REGION.amazonaws.
com/streamlit-app:latest

# 5. Push to ECR
docker push $AWS_ACCOUNT_ID.dkr.ecr.$REGION.amazonaws.com/streamlit-
app:latest
```

Both repositories now contain the latest image for their respective applications. In the next section, you'll deploy these images with AWS ECS Fargate behind an HTTPS endpoint.

Preparing the Ground for ECS (Elastic Container Service)

Activate the ECS Service-Linked Role

Navigate to **IAM ➤ Roles** and search for AWSServiceRoleForECS, and click it; that's it. This process can seem unnecessary, but that single click is enough to finish activating the role.

Without the service-linked role, creating an ECS cluster can fail with a 400 Invalid Request because ECS cannot bind the service to the required permissions.

The IAM user that is doing this deployment needs to have this role added to them.

Create the Required Security Groups

Navigate to **EC2 ➤ Network & Security ➤ Security Groups** and click `Create security group`. We're going to create one security group for the Application Load Balancer (ALB) and one for the ECS services.

Starting with the security group for the ALB

- Give it a descriptive name like `alb-sg`.

- Add a description like **Security Group for the Application Load Balancer**.

- Check that the VPC is the right one.

- **Inbound rules:** Click `Add rule` and configure it like Table 10-3. See also Figure 10-5.

Table 10-3. *Configuring inbound rules for ALB*

Type	Protocol	Port	Source	Description
HTTP	TCP	80	`<YOUR_IP>/32` (e.g., `87.154.111.11/32`)	Allow HTTP traffic *only* from your workstation

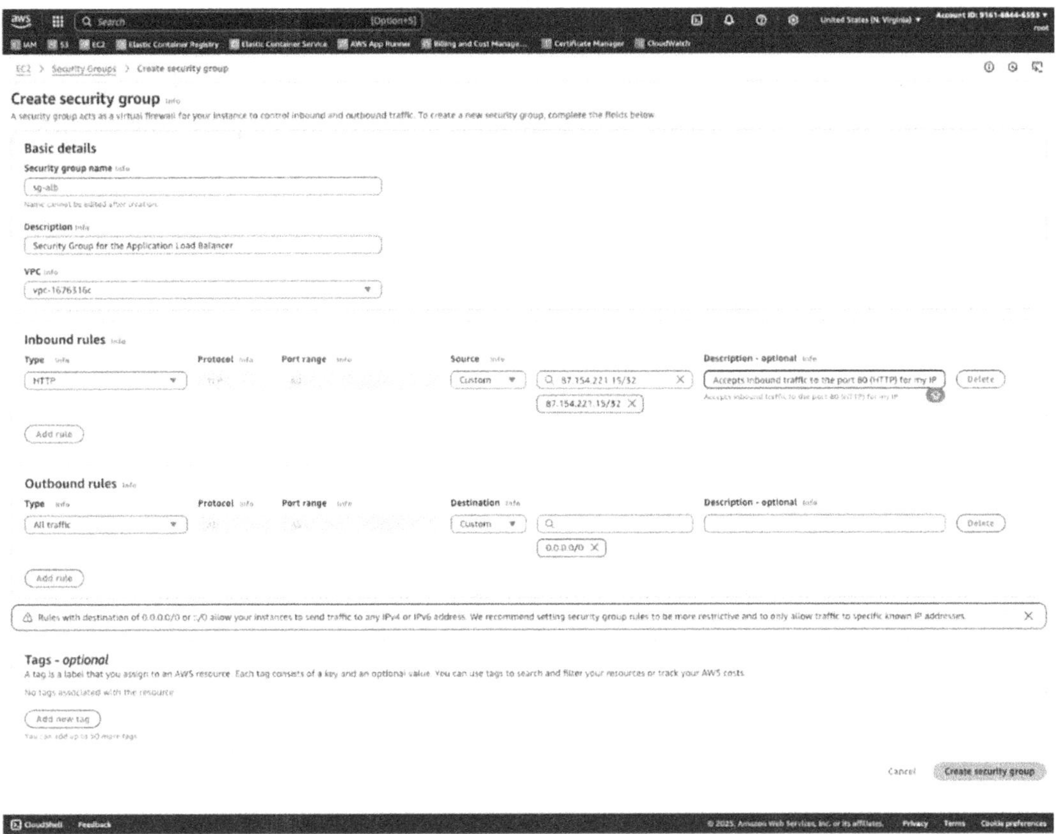

Figure 10-5. *Setting up the ALB security group*

Now create the security group for ECS:

- Give it a descriptive name like: `ecs-sg-tasks`.

- **Description: ECS security group tasks**.

- Check that the VPC is the right one.

- **Inbound rules:** Click `Add rule` and add a rule for Streamlit;
  configure it like Table 10-4.

Table 10-4. *Configuring inbound rules for ECS services*

#	Type	Protocol	Port	Source	Description
1	Custom TCP	TCP	8501	`alb-sg` (select from the SG list)	*Streamlit* – traffic allowed only from ALB
2	Custom TCP	TCP	5000	`alb-sg`	*Flask* – traffic allowed only from ALB

Note that when choosing the Source, instead of an IP, you want to select the **security group alb-sg** so that **only** the Application Load Balancer (ALB) can access the ECS containers; see Figure 10-6.

Then click `Create security` group at the bottom right of the page.

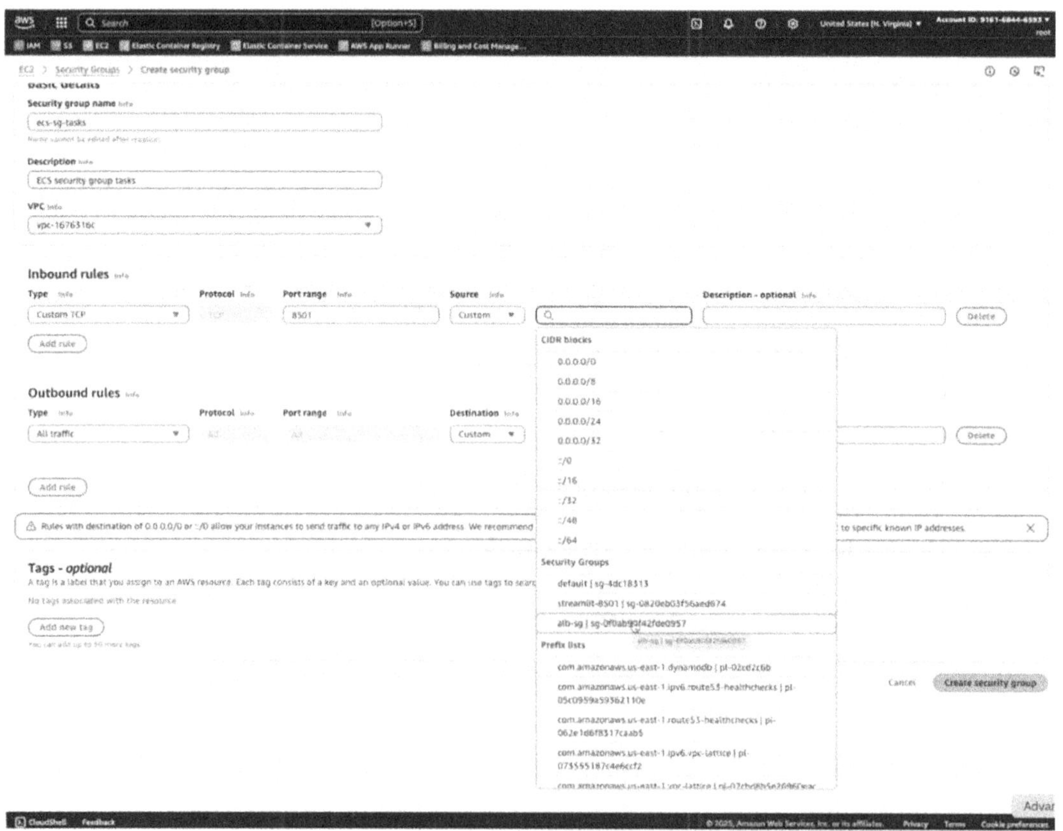

Figure 10-6. *Setting up the ECS security group*

Create Target Groups

Navigate to **EC2 ➤ Load Balancing ➤ Target Groups** and click `Create target group`. You need two target groups: one for **Streamlit** and one for **Flask**. Starting with Streamlit. See Figure 10-7.

Target Group for Streamlit

1. **Choose the target type: IP addresses**

2. Follow configuration from Table 10-5.

Table 10-5. *Target group configuration for Streamlit*

Field	Value
Target group name	streamlit-tg
Protocol	HTTP
Port	8501
IP address type	IPv4 (default)
Protocol version	HTTP1 (default)

3. Click **Next** to reach the **Register targets** screen.

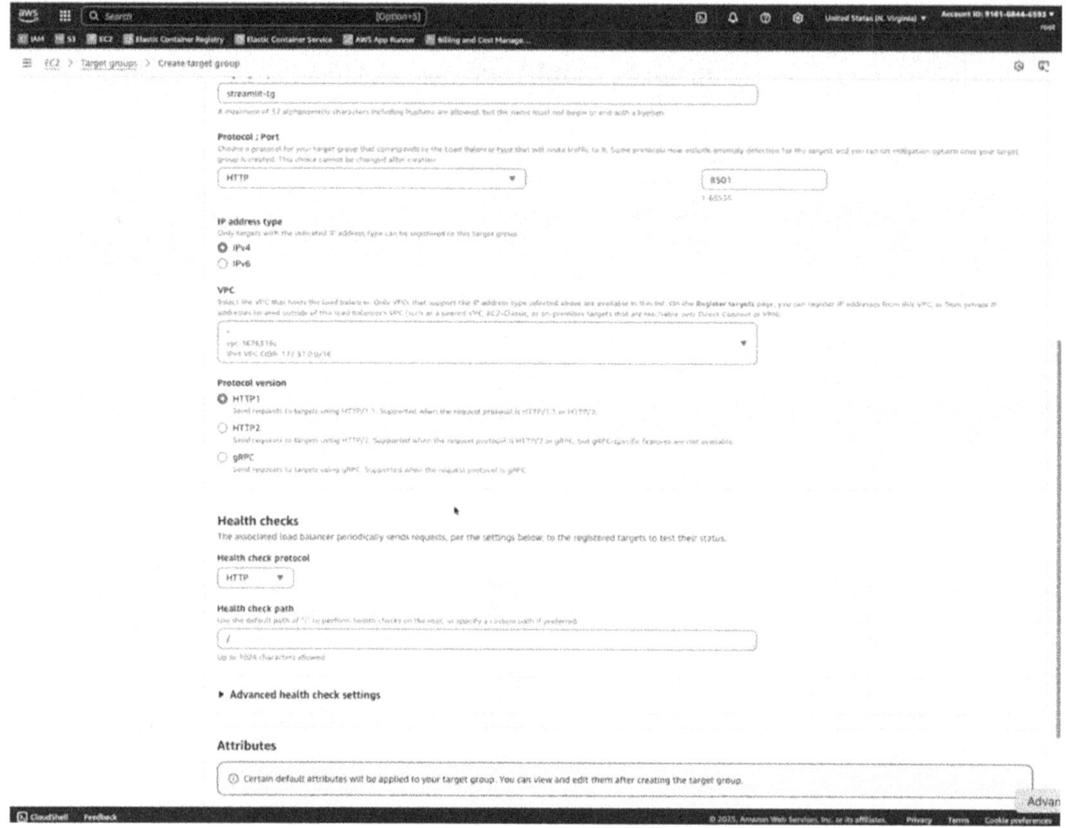

Figure 10-7. *Create target group*

4. **Remove any pre-filled IP**

    A placeholder IP may appear; click **Remove** so the list is empty;
    see Figure 10-8.

5. Click ***Create target group***.

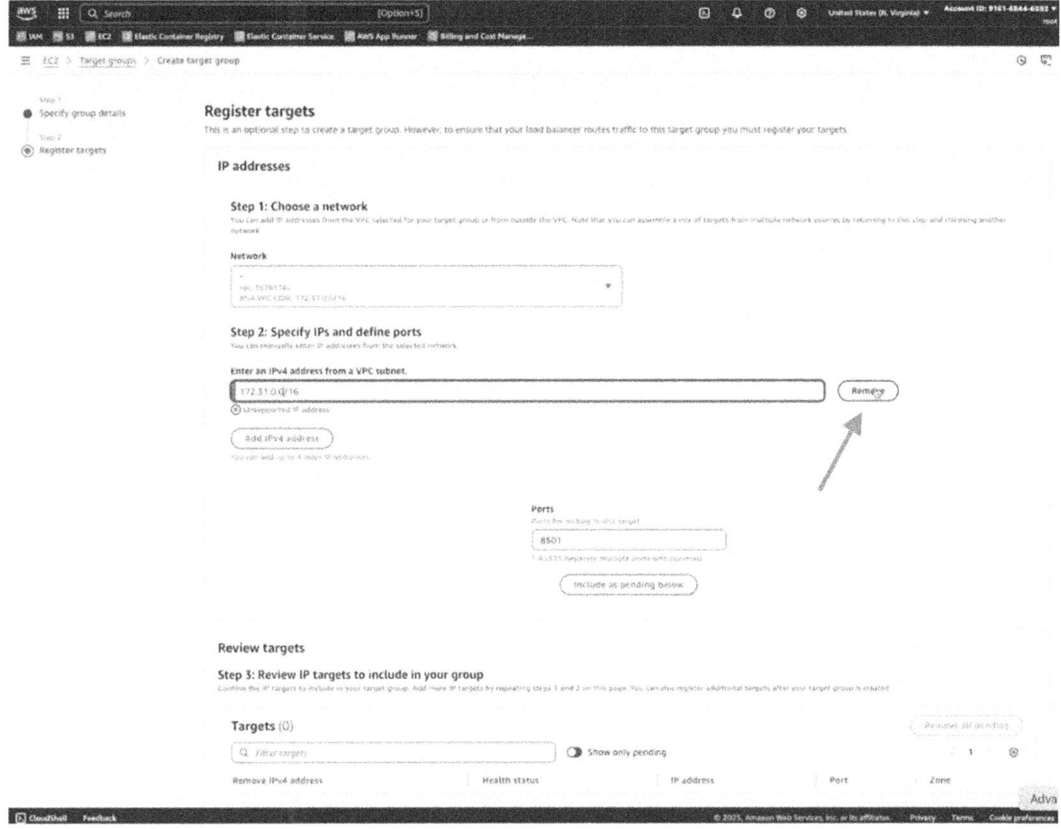

Figure 10-8. *On target group register targets page*

Why No Static IPs?

In a serverless (Fargate) environment, AWS dynamically attaches and detaches task Elastic Network Interface (ENI). Leaving the target list empty lets the Application Load Balancer register those IPs automatically.

Target Group for Flask

Repeat the same steps above with Flask-specific values; see Table 10-6.

Table 10-6. *Target group configuration for Flask*

Field	Value
Target group name	`flask-tg`
Protocol	`HTTP`
Port	`5000`
IP address type	`IPv4`
Protocol version	`HTTP1`

Creating the Application Load Balancer (ALB)

Open the EC2 console, expand **Load Balancing**, choose **Load Balancers**, and click **Create load balancer**. Select **Application Load Balancer**, then work through the wizard as follows:

1. **Basic details:** Give the ALB a clear name such as **ecs-fargate-alb**. Keep the **IP address type** on **IPv4** and set the **scheme** to **Internet-facing** so the service is publicly reachable.

2. **Network mapping:** Select the same VPC you have used throughout this chapter. When the list of subnets appears, enable every Availability Zone. Turning on all AZs (see Figure 10-9) doesn't raise the hourly ALB charge, but it guarantees that tasks scheduled in any zone can register with the load balancer. The trade-off is that cross-AZ data transfer or task placement can add a small cost and a little latency, yet the resilience it provides is worth it.

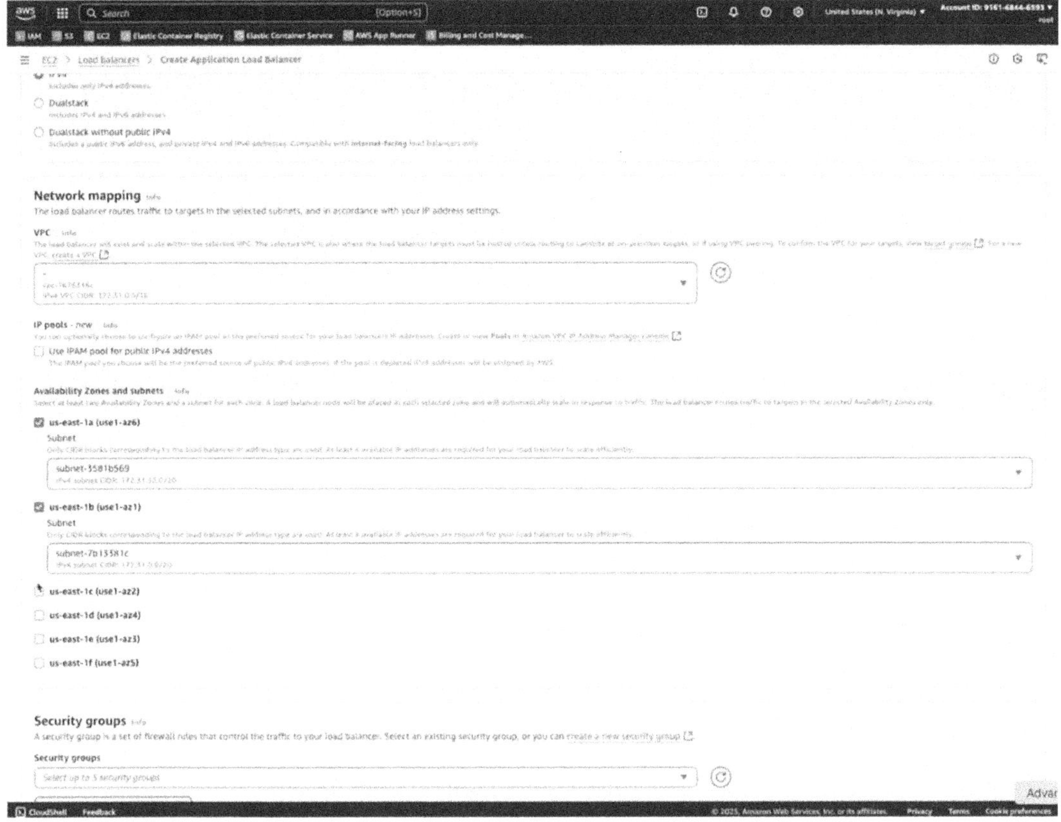

Figure 10-9. *Enable all AZs*

3. **Security groups:** Attach the group you created earlier for the load balancer, **alb-sg**. This group already allows inbound HTTP traffic on port 80 from your workstation (or whichever IP range you specified).

4. **Listeners and default action:** Start with a single listener: **HTTP** on port **80**. Set its default action to **forward to flask-tg**. We will refine the rules after the ALB is active, adding host-based routing so requests for streamlit.lb-webserver.pro are sent to **streamlit-tg**.

Review the settings and click **Create load balancer**. AWS now provisions the ALB, attaches it to every selected subnet, and applies the security group. When the new load balancer shows a status of **active**, you are ready to add the routing rules and lock down the default DNS name in the next step.

Deny Access via the Load Balancer's Default DNS Name

We want users to reach the applications only through your own domain (`lb-webserver.pro`). To block the automatically generated DNS name that AWS assigns to every load balancer:

1. In the EC2 console, return to **Load Balancing ➤ Load Balancers**, select **ecs-fargate-alb**, and open the **Listeners and rules** tab at the bottom. Click the single listener, **HTTP : 80**; see Figure 10-10.

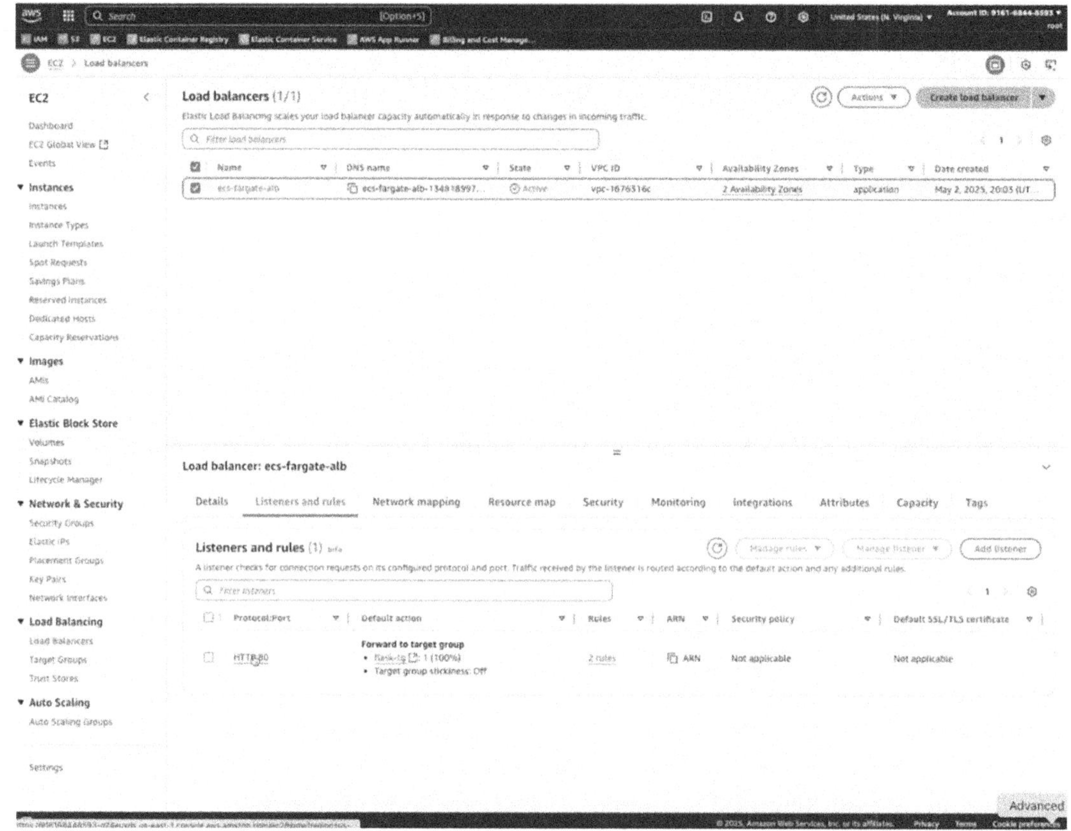

Figure 10-10. *Adjust Listeners and rules*

2. The rules table lists two entries: ***Default*** and ***streamlit***. Tick the box next to ***Default***, then choose ***Actions ➤ Edit rule*** (Figure 10-11).

234

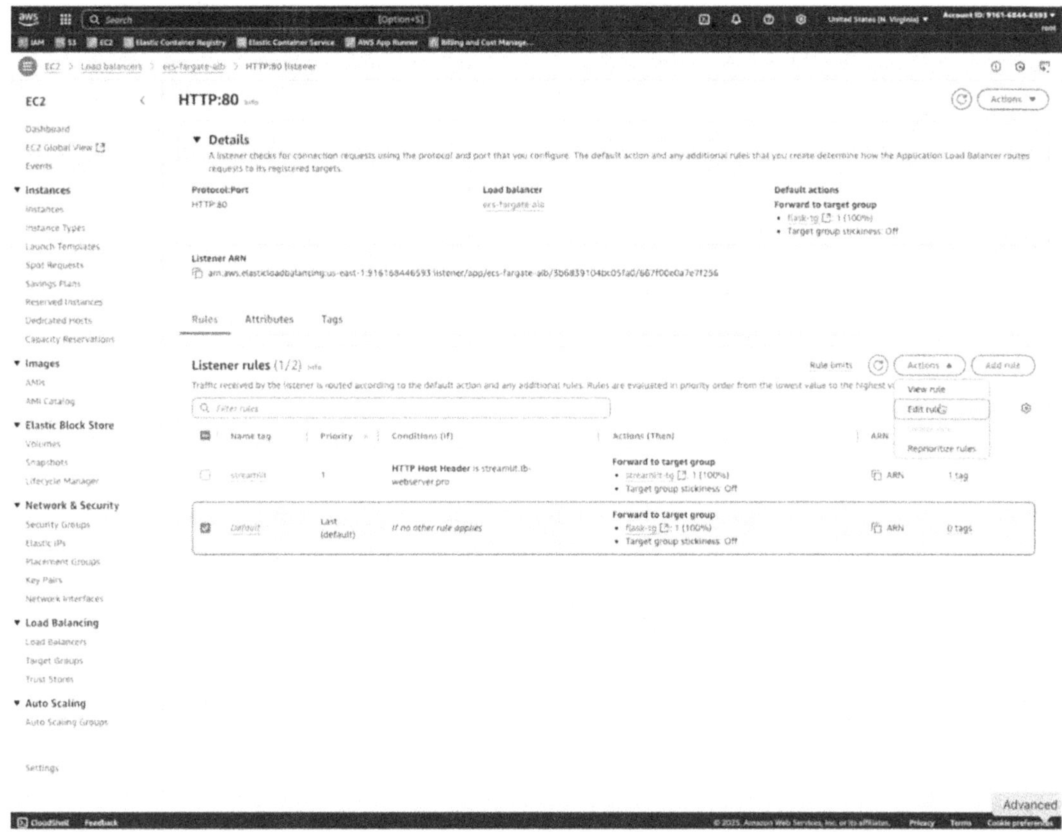

Figure 10-11. *Adjust the default*

3. A rule editor opens. Under **Default actions**, change the action type to **Return fixed response**.

    - Set **Response code** to **403**.

    - In **Response body**, add a friendly hint—for example:

    "Access via this DNS is denied. Please use the official domain."

4. Click ***Save changes*** (lower right). From now on, anyone who tries the ALB's DNS endpoint will receive a 403 message, while your custom domains will keep working as intended; see Figure 10-12.

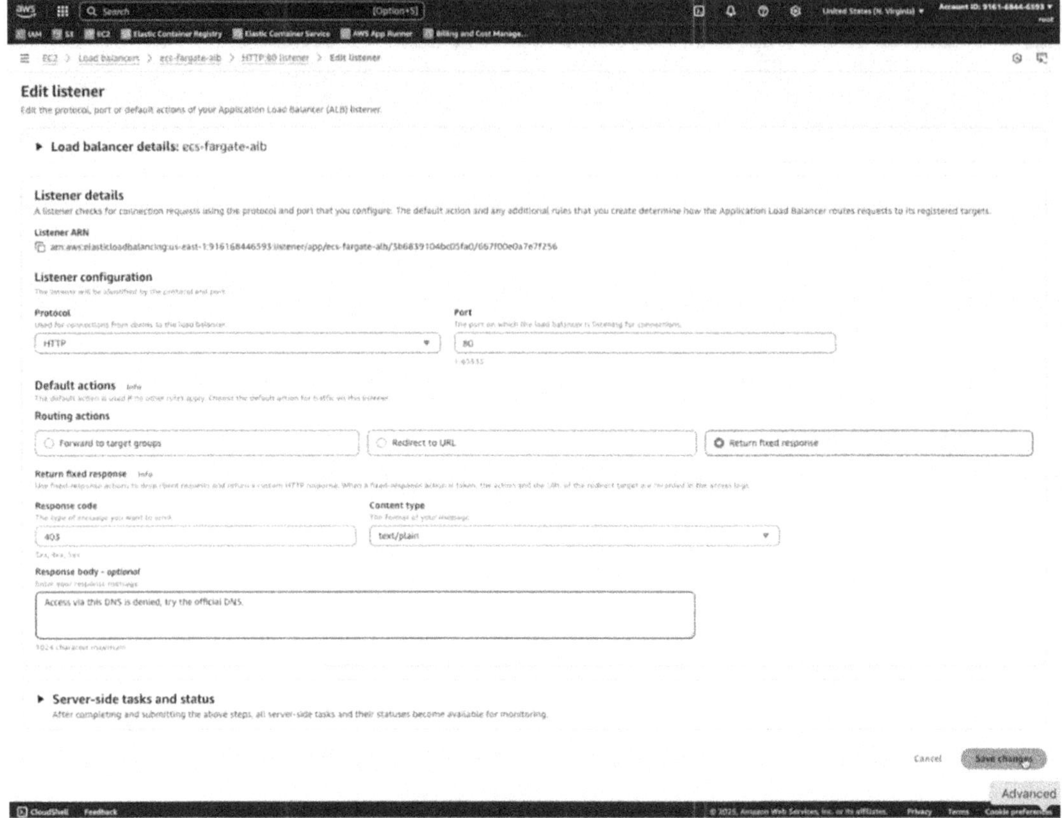

Figure 10-12. *Deny access via ALB's DN*

Note   AWS's Application Load Balancer includes AWS Shield Standard by default, which provides built-in DDoS protection at no extra cost; no additional configuration is needed.

Adjust Listener and Rules to Access Apps with Subdomains in HTTP

With the default DNS now blocked, the next step is to direct traffic for each custom subdomain to its corresponding target group.

1. **Edit the existing Streamlit rule**

   Still on the **Listeners and rules** tab for **ecs-fargate-alb**, tick the
   row labeled **streamlit** and choose **Actions ➤ Edit rule** (refer back
   to Figure 10-11).

   - Click **Add condition**, select **Host header**, and enter the full
     subdomain:

     - streamlit.lb-webserver.pro

     Click **Confirm** (Figure 10-13).

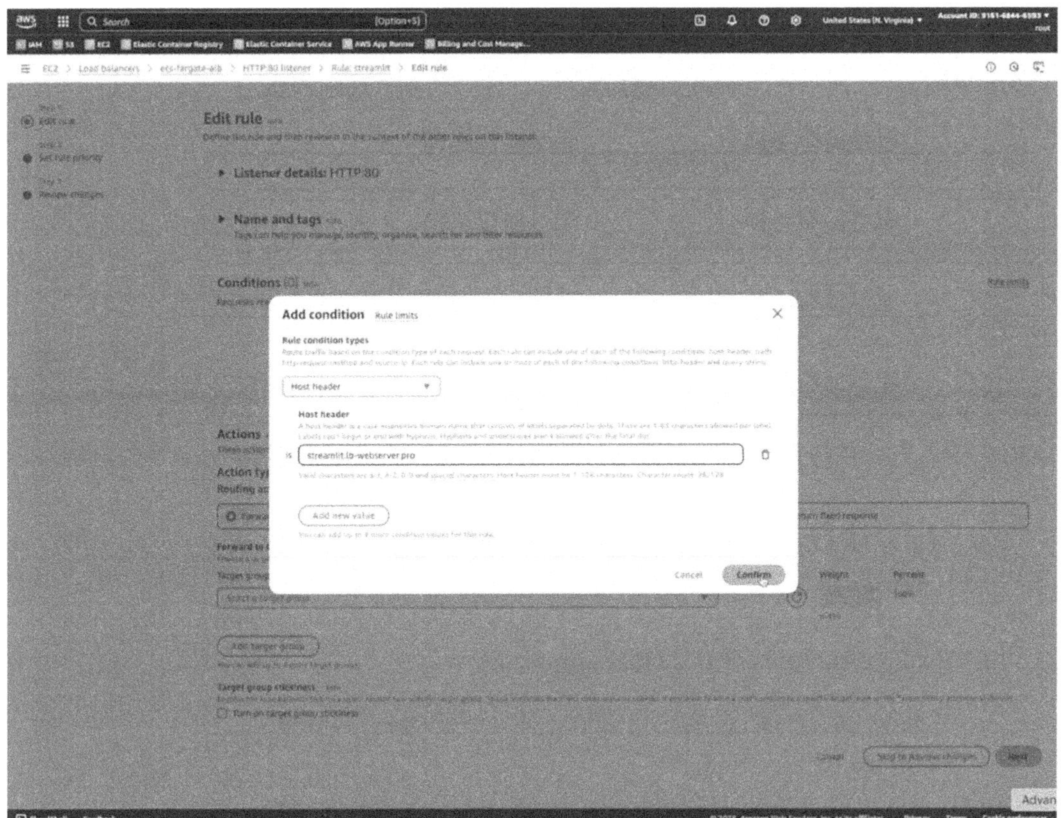

Figure 10-13. *Adjust the Streamlit route*

   - Under **Actions**, change the action to **Forward to target groups**
     and pick **streamlit-tg** (see Figure 10-14).

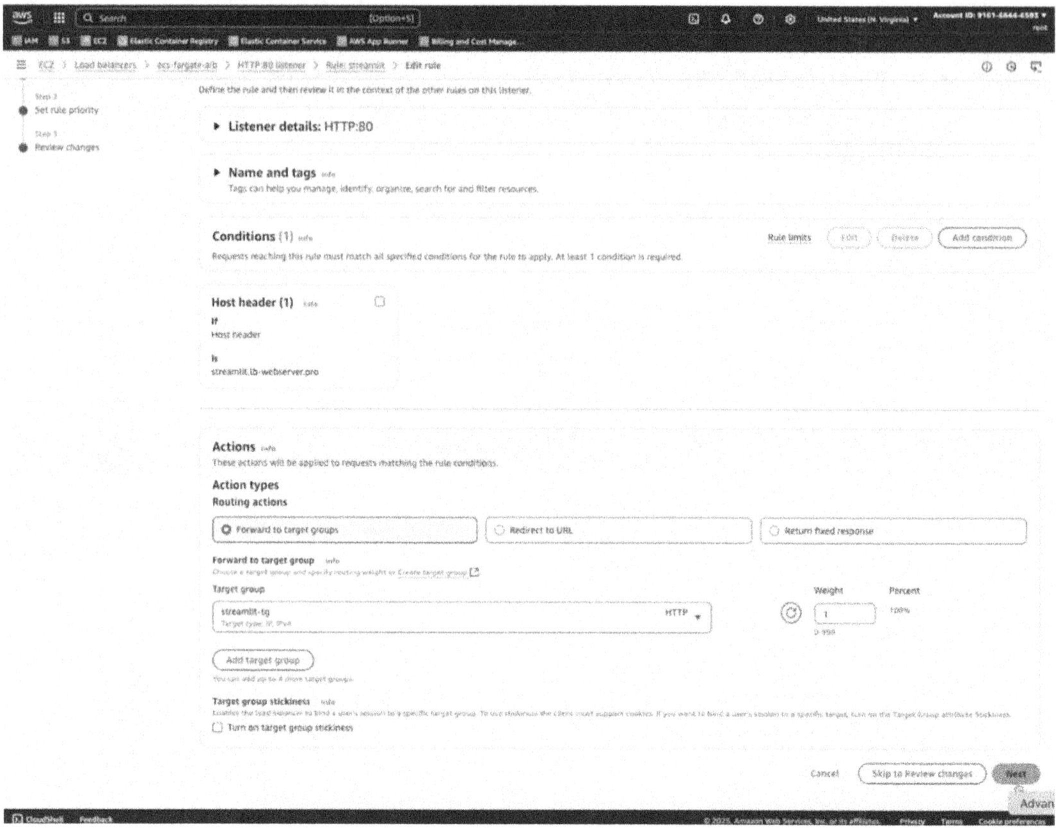

Figure 10-14. *Adjust the Streamlit actions to forward to target group streamlit-tg.*

- Click **Next**, review the summary, click **Next** again, then **Save changes**.

2. **Add a new rule for the Flask application**

   You are returned to the rules list. Click **Add rule** to open a blank editor:

   - **Condition:** Choose **Host header** and enter

     - `flask.lb-webserver.pro`

   - **Action:** Set to **Forward to target groups** and select **flask-tg**.

   - Step through **Next ➤ Next** and finish with **Save changes**.

When you return to the rules table, you should see three entries, as shown in
Table 10-7.

Table 10-7. Listeners and rules

Priority	Condition	Action
1	Host = `streamlit.lb-webserver.pro`	Forward to `streamlit-tg`
2	Host = `flask.lb-webserver.pro`	Forward to `flask-tg`
Default	—	Fixed response 403

Figure 10-15 illustrates this final configuration.

At this point, HTTP requests for each subdomain are routed to the correct ECS tasks,
while any attempt to use the ALB's native DNS name receives a 403 response.

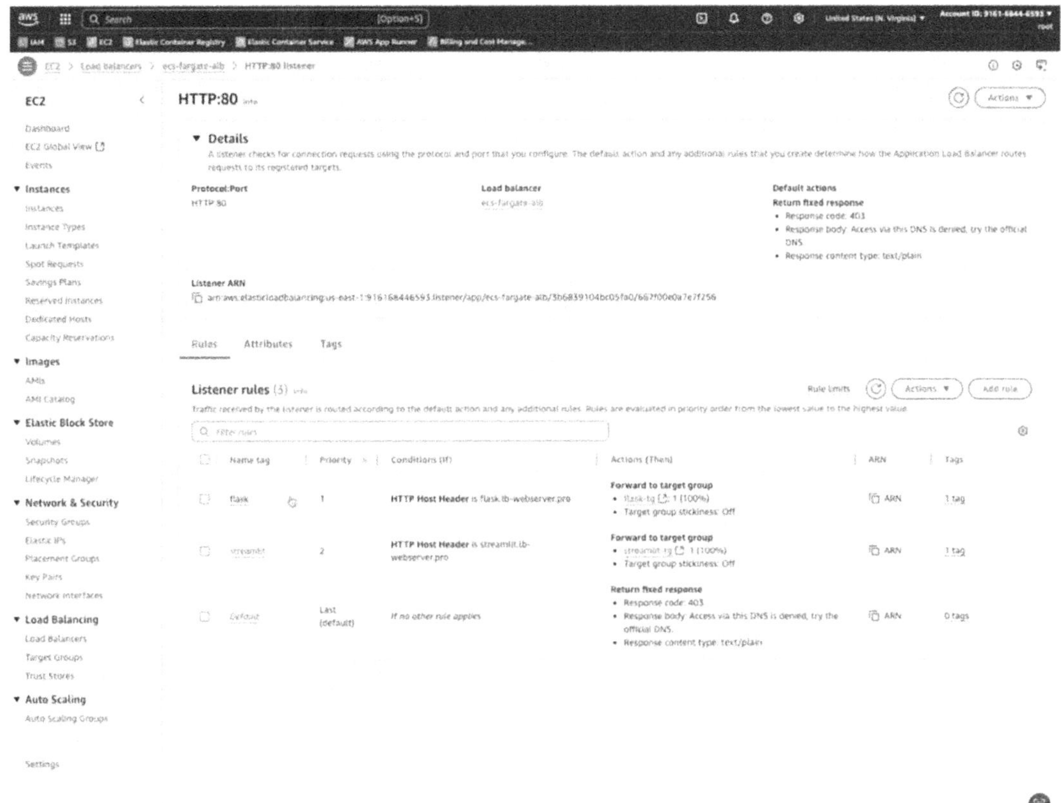

Figure 10-15. Listener rules' final result

Deployment with ECS and Fargate

With the load-balancing layer complete, the remaining job is to spin up the containers on AWS Fargate and let the ALB register them automatically.

Create an ECS Cluster

1. Open **Amazon Elastic Container Service ➤ Clusters** and click **Create cluster**.

2. Give the cluster a descriptive name such as **ecs-fargate-cluster**.

3. Under **Infrastructure**, keep **AWS Fargate (serverless)** selected and leave **Amazon EC2 instances** unticked.

4. Click *Create*. The new cluster appears almost instantly; see Figure 10-16.

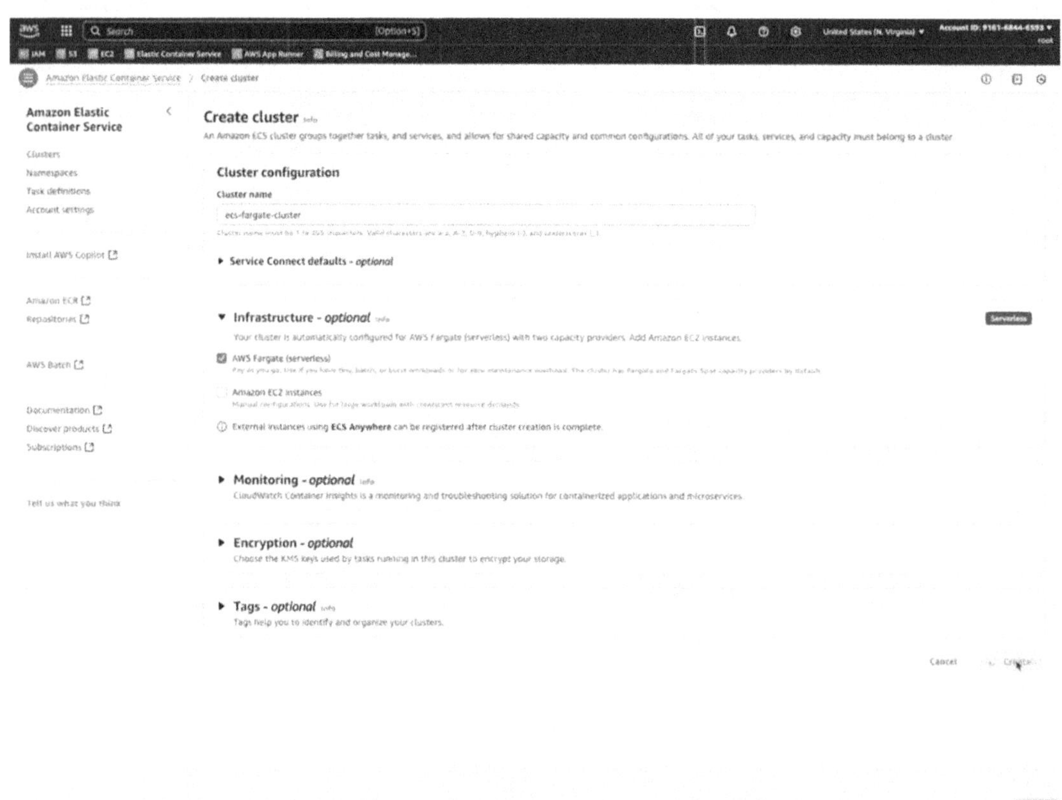

Figure 10-16. *Cluster created*

Create Task Definitions

A *task definition* is to ECS what a `docker-compose.yml` file is to Docker Compose: it declares which image to run, what resources it needs, and which ports it exposes. You will create one task definition for each application.

Task Definition for *Streamlit*

1. Navigate to **ECS ➤ Task definitions** and choose **Create new task definition**.

2. Fill in the form as follows:

   Basic settings

   - Family name: **streamlit-app**

3. **Infrastructure requirements**

   - **Launch type: AWS Fargate**

   - **OS/Architecture: Linux / x86-64** (default)

   - **Task size: 1 vCPU** and **3 GB** of memory

4. **Roles**

   - **Task role:** - (leave blank)

   - **Task execution role: Create new role** (accept the default)

5. **Container 1**

   - Name: **streamlit-app**

   - **Image URI:** `916168446593.dkr.ecr.us-east-1.amazonaws.com/streamlit-app:latest` (same image that you pushed to ECR)

   - **Essential container: Yes**

   - **Container port: 8501 / TCP**

6. Scroll to the bottom and click **Create**.

   The console confirms the task definition; see Figure 10-17.

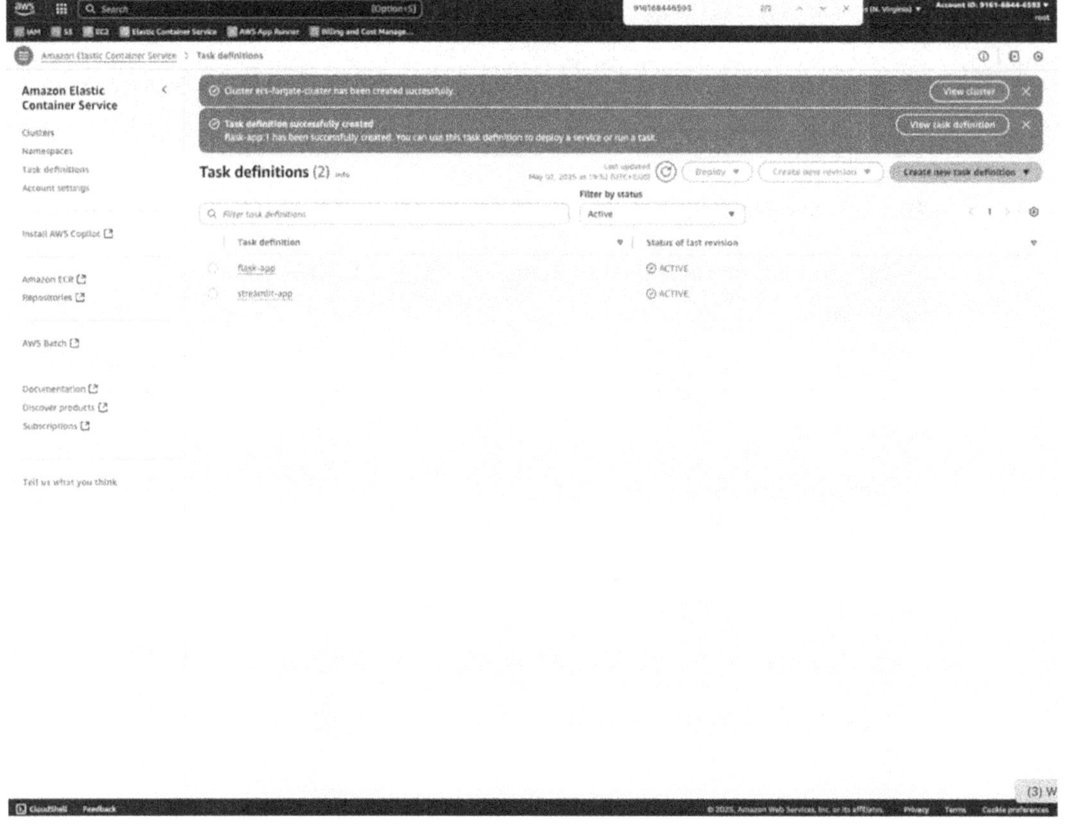

Figure 10-17. *Task definition created*

Task Definition for *Flask*

Repeat the process, adjusting only the fields that differ:

- **Family name: `flask-app`**

- **Task size: 2 vCPU** and **4 GB** of memory (Flask benefits from a little more headroom)

- **Task role and Task execution role:** Select the **`ecsTaskExecutionRole`** created during the Streamlit task definition

- **Image URI:** `916168446593.dkr.ecr.us-east-1.amazonaws.com/flask-app:latest`

- **Container port: 5000 / TCP**

Leave every other setting at its default and click **Create**.

You now have two task definitions ready to be launched as Fargate services inside the cluster. In the next section, we will deploy those services, attach them to the target groups, and watch the ALB route traffic to the correct container.

Create ECS Services

With two task definitions in place, the final step is to launch them as **Fargate services** inside the cluster and wire each one to the Application Load Balancer.

1) **Flask service**

   1. Open **ECS ➤ Clusters**, click **ecs-fargate-cluster**, switch to the **Services** tab, and choose **Create**.

   2. **Service details:** In the first drop-down, select the task definition **flask-app**; see Figure 10-18. Leave the remaining options at their defaults and continue to **Networking**.

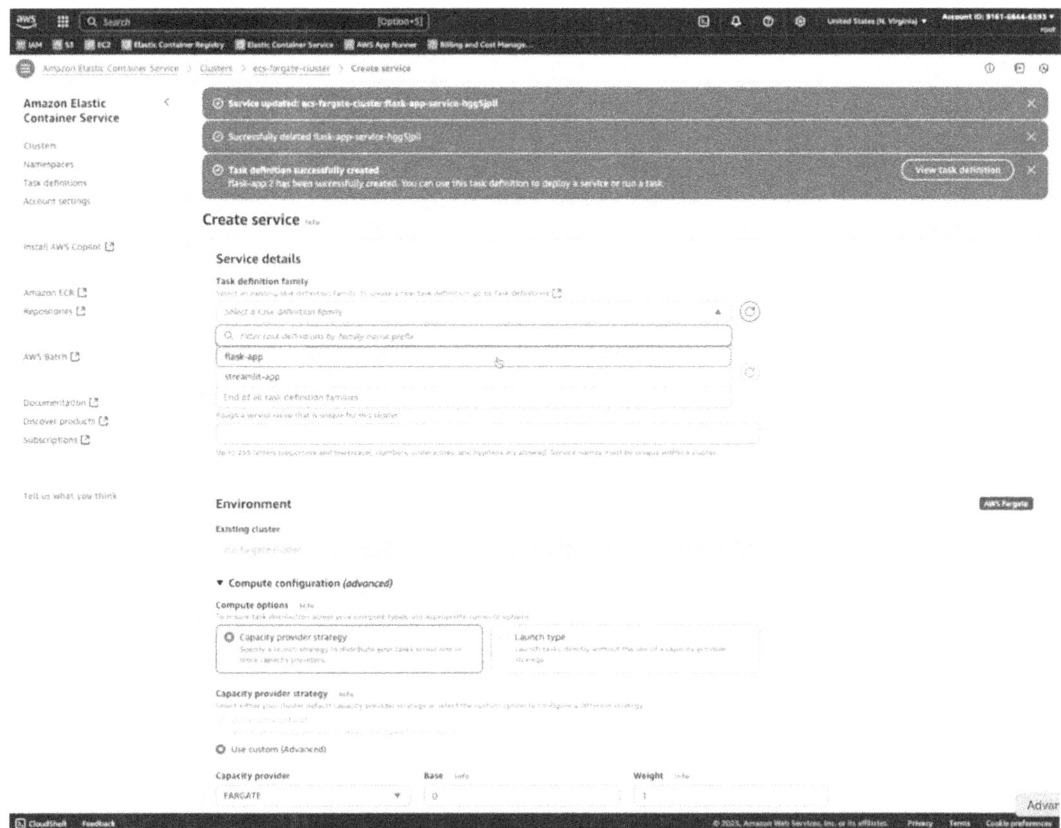

Figure 10-18. *Create service*

3. **Networking:** Verify the correct VPC is selected and that all subnets, one in each Availability Zone, are ticked. In the **Security group** field, deselect the default SG and choose **ecs-sg-tasks**; see Figure 10-19.

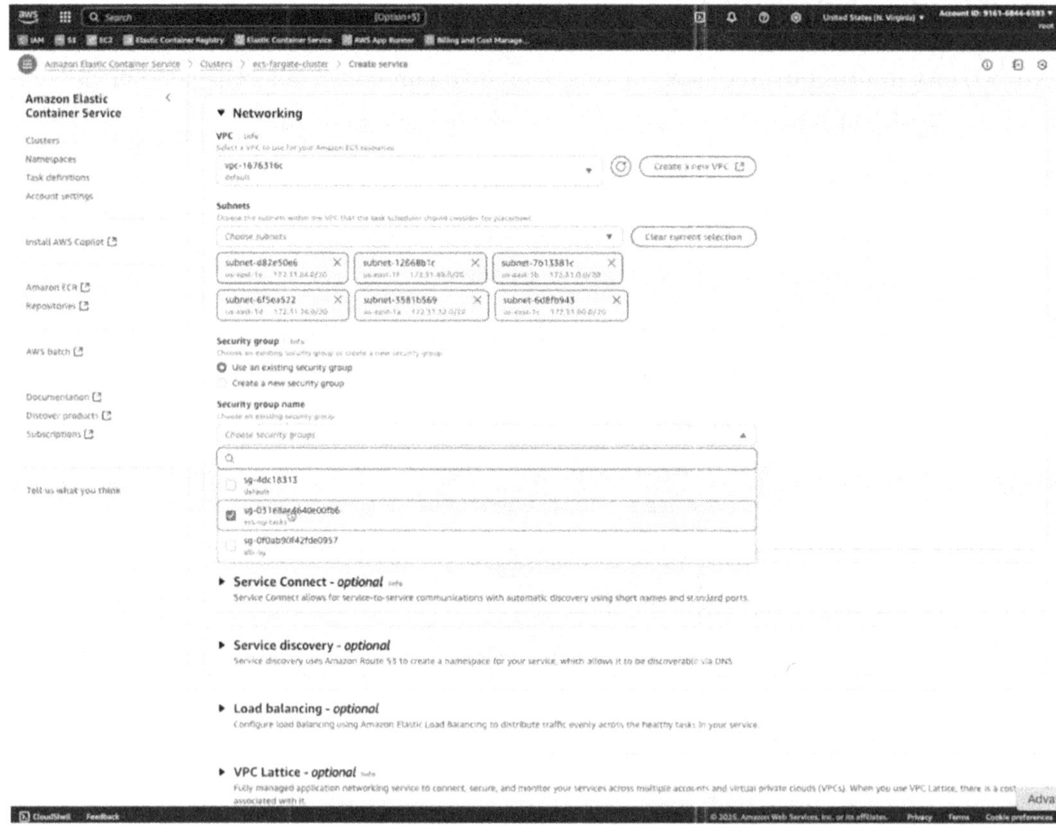

Figure 10-19. *On Networking, choose the correct security group*

4. **Load balancing:** Expand the section and tick **Use load balancing**; see Figure 10-20.

   - **Type:** *Application Load Balancer.*

   - **Container:** Should auto-populate as **flask-app 5000:5000**.

   - **Application Load Balancer:** Choose an existing load balancer, **ecs-fargate-alb**.

- **Listener:** Choose the existing **HTTP : 80** listener.

- **Target group:** Also choose an existing target group,
  flask-tg.

5. Review the summary and click **Create**. The service deploys;
    once its status turns green, the Flask app is live behind
    the ALB.

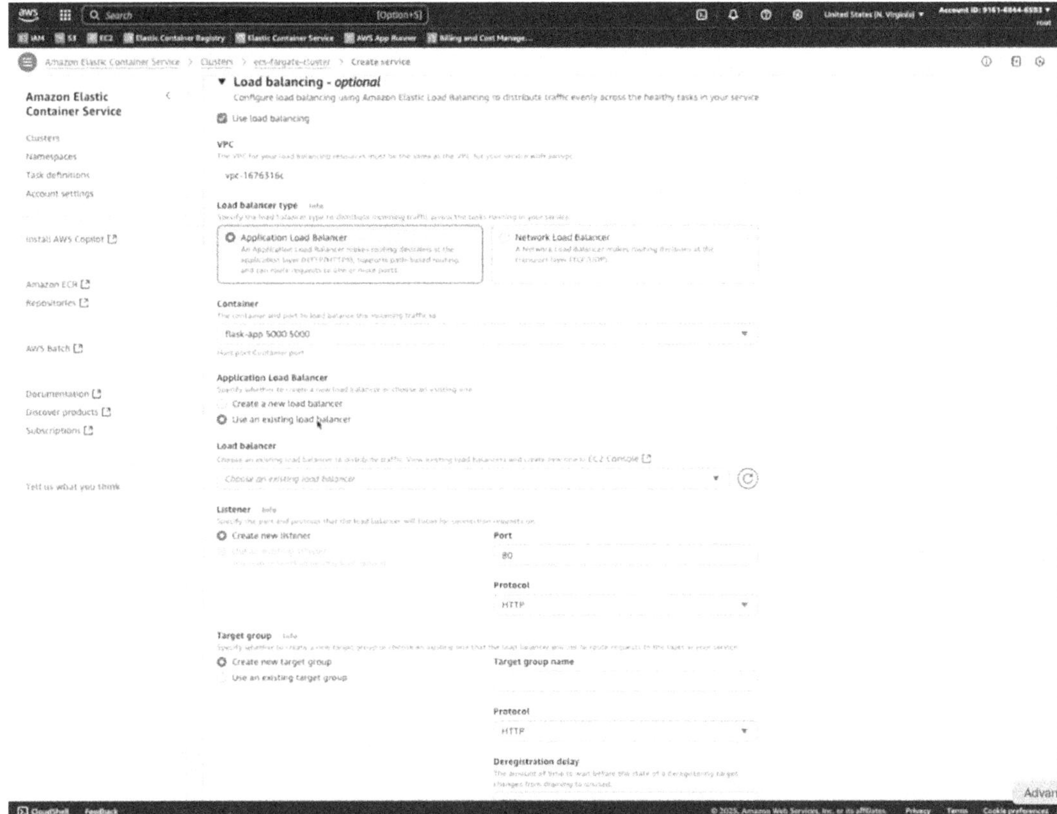

Figure 10-20. *Use load balancing*

2) **Streamlit service**

Repeat the workflow to add the Streamlit task. Navigate to
Amazon Elastic Container Service ➤ Clusters, and choose your
cluster `ecs-fargate-cluster`.

1. Back on the **Services** tab, click **Create** again.

2. **Service details:** Choose the **streamlit-app** task definition.

3. **Networking:** Switch the security group to **ecs-sg-tasks** just as before.

4. **Load balancing:** Enable **Use load balancing** and confirm:

    - **Container: streamlit-app 8501:8501** (should be pre-selected)

    - **Application Load Balancer: ecs-fargate-alb** (use an existing load balancer)

    - **Listener: HTTP : 80** (use an existing listener)

    - **Target group: streamlit-tg** (use an existing target group)

5. Click **Create**. When the service indicator turns green (see Figure 10-21), Streamlit is available at its subdomain.

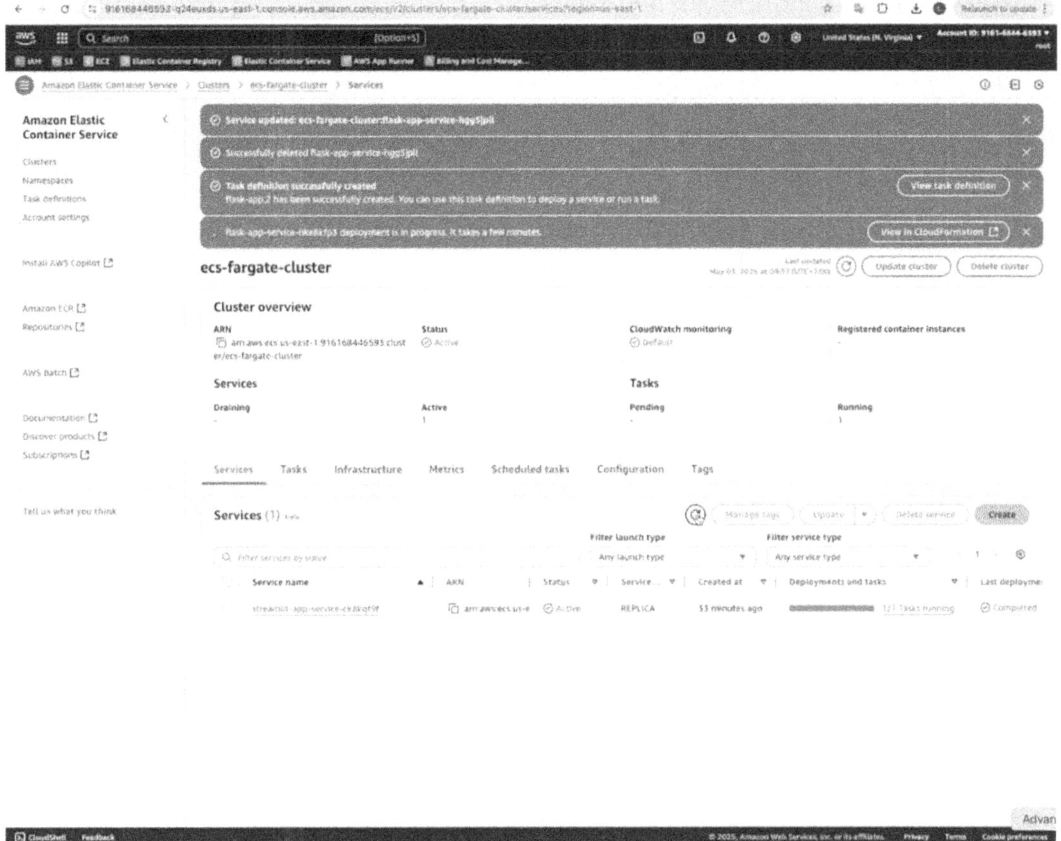

Figure 10-21. *Services are deploying*

Monitoring Logs with CloudWatch and the ECS Console

After the service is running, everything the container prints, startup messages, request details, warnings, and errors, is streamed to **Amazon CloudWatch Logs**. You can inspect those messages in two places: CloudWatch itself or the Logs tab inside the ECS console.

Every line your application would normally write to the terminal (standard output and standard error) is captured by the ECS agent and forwarded to CloudWatch. This means you can troubleshoot the container without SSH-ing into a host.

To access these logs through CloudWatch:

1. Navigate to **CloudWatch ➤ Logs ➤ Log groups**, and be sure you're in the same AWS region as the cluster.

2. Locate the log group named something like `ecs/flaskapp`; the pattern is `/ecs/<task-definition-family>`.

3. Click the group, then pick the most recent entry under `Log streams`.

You will see the live stream of application messages just as if you were watching them in a terminal; see Figure 10-22.

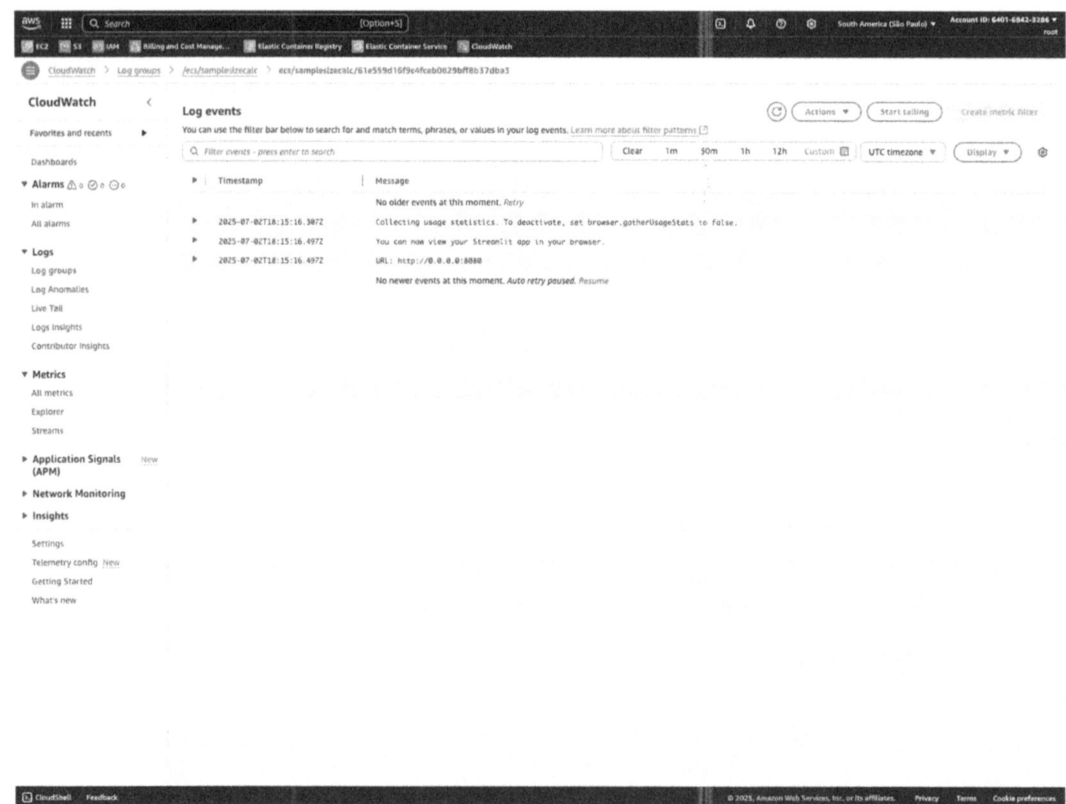

Figure 10-22. *Logs from CloudWatch*

To access these logs through ECS:

1. Navigate to **ECS ➤ Clusters**, choose your cluster, and select the **Services** tab.

2. Click the service you want to inspect, then switch to the **Logs** tab.

The console embeds the same CloudWatch stream, so you can debug without leaving the ECS view.

Adjust CNAME in Namecheap

In the meantime that the services are deploying, we need to adjust the DNS on Namecheap so that the domain and subdomains work properly with the Application Load Balancer that is handling the ECS + Fargate services.

1. **Log in to Namecheap** and open the **Advanced DNS** tab for `lb-webserver.pro`.

2. For each host record: @, `flask`, and `streamlit`, change the **Type** from **A Record** to **CNAME Record**.

3. In the **Value** field, paste the DNS name of the Application Load Balancer (find it in AWS under **EC2 ➤ Load balancers**).

4. Save your changes; see Figure 10-23.

Once DNS propagates, the root domain and both subdomains will resolve to the ALB, which in turn routes traffic to the correct ECS service.

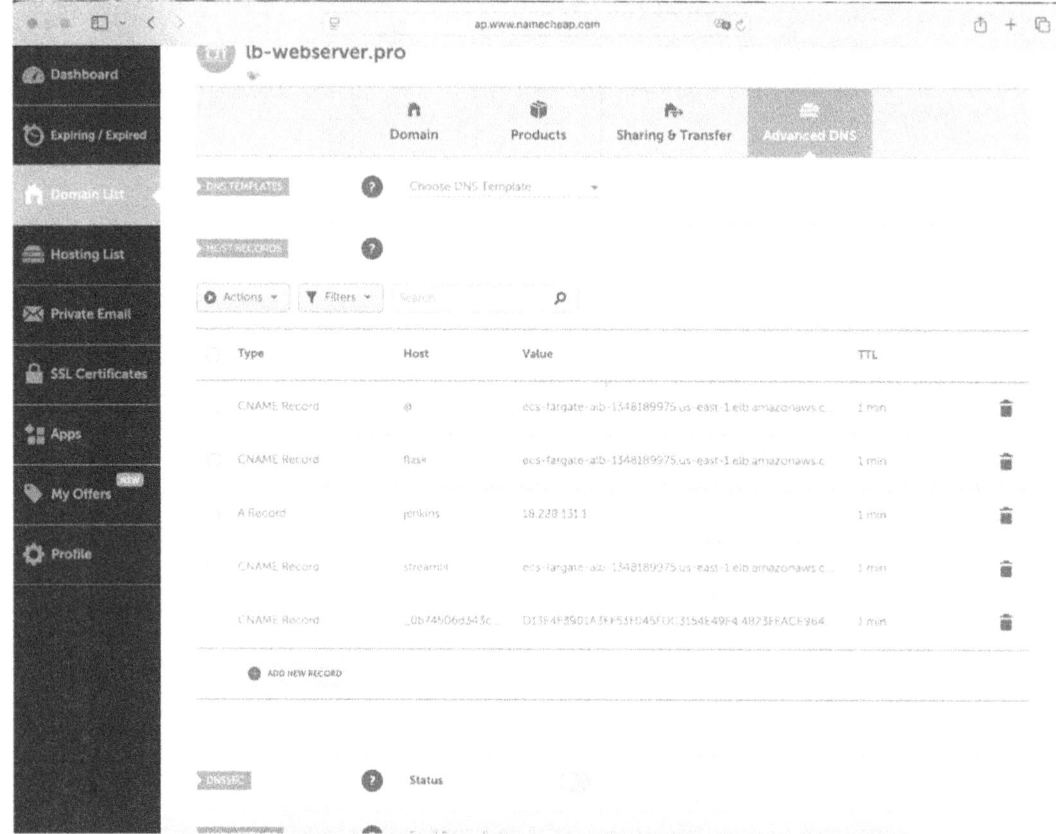

Figure 10-23. *Adjust Advanced DNS on Namecheap*

Enabling HTTPS with Your Own SSL Certificate

The Application Load Balancer is already serving HTTP traffic, but production
sites should encrypt everything with TLS. Because you purchased a certificate from
Namecheap in Chapter 4, section "Set Up the SSL Certificates," you can simply import
that certificate into **AWS Certificate Manager (ACM)** and attach it to the existing ALB
listener.

Regional Reminder

ACM certificates are regional resources. Import the certificate in the same AWS
region where the ALB resides; otherwise it will not appear in the listener's drop-
down list.

Import the Certificate into ACM

1. In the AWS console search bar, type **Certificate Manager** and open the service.

2. Click **Import a certificate**; see Figure 10-24.

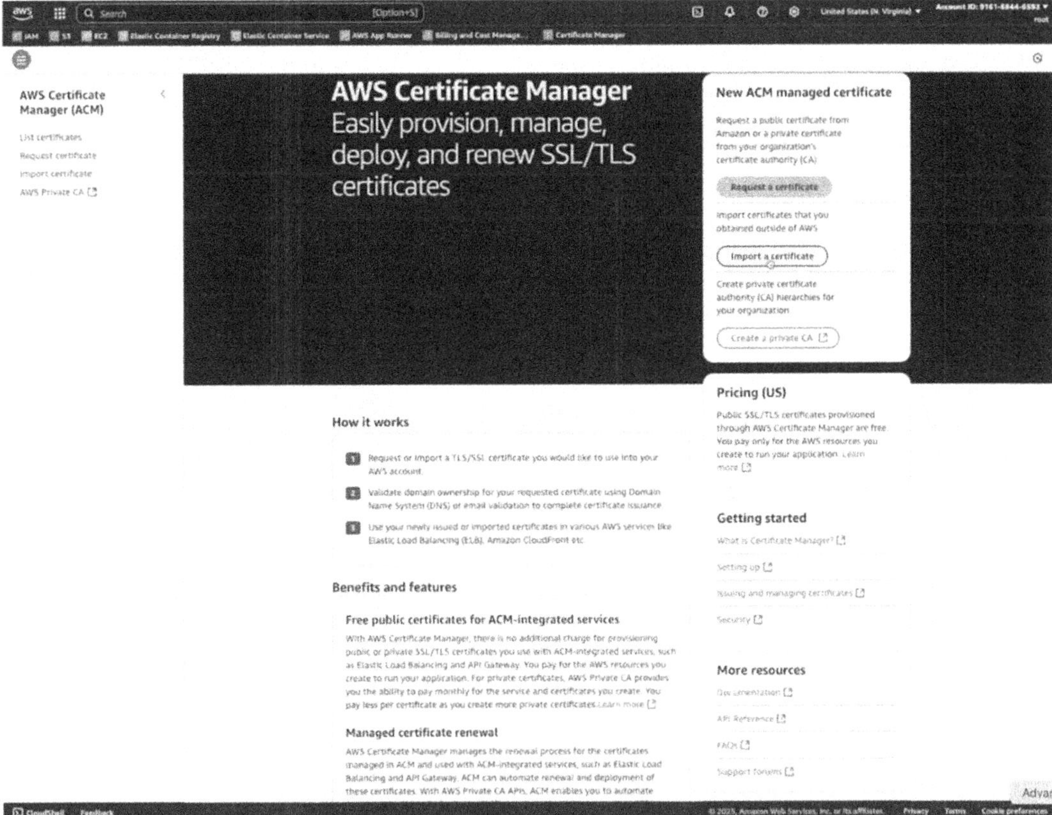

Figure 10-24. *Import certificate into AWS Certificate Manager (ACM)*

3. Fill the three text areas with the contents of the files you downloaded from Namecheap:

   - **Certificate body:** Here you paste the contents of your_
     domain.crt.

   - **Certificate private key:** Here you paste the contents of
     private.key.

- **Certificate chain - optional:** Here you paste the contents of
  `your_domain.ca-bundle`.

Figure 10-25 shows the form.

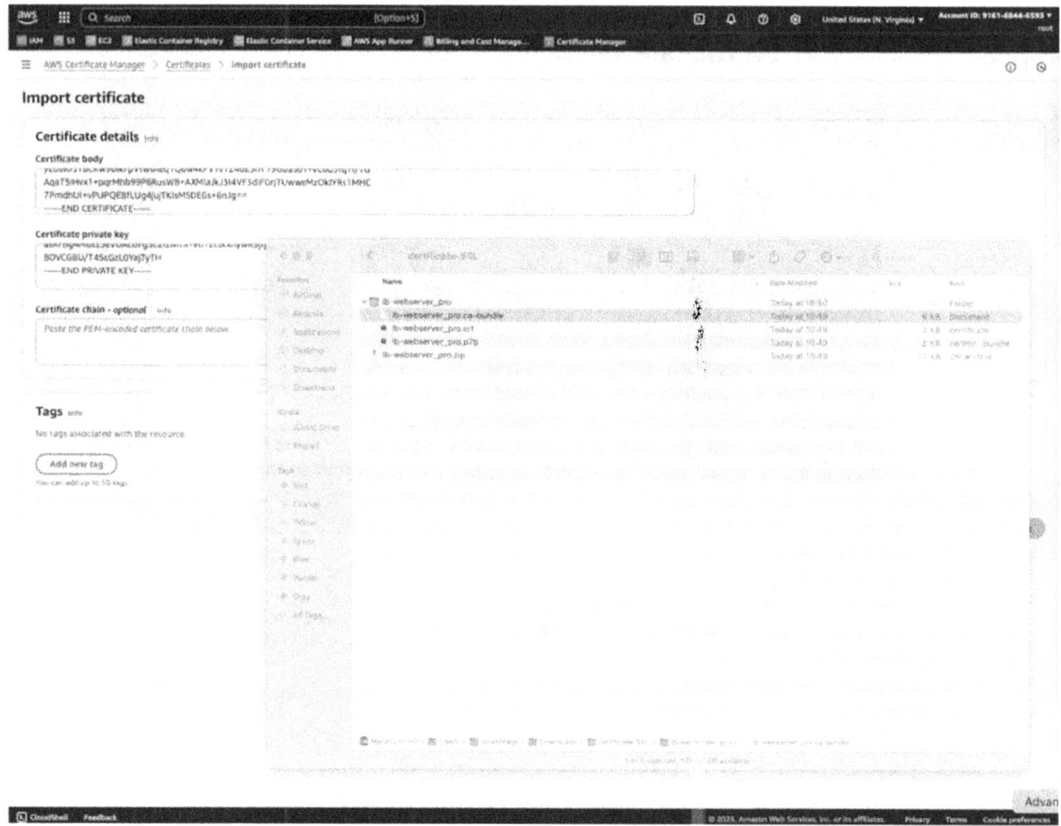

Figure 10-25. *Paste the certificate content to ACM*

4. Click **Import certificate**. ACM validates the key pair and adds the
   certificate to your account. Once the status is **Issued**, it is ready
   for use.

With the certificate safely stored in ACM, the next step is to add an HTTPS listener
to the ALB and associate this certificate, completing the secure end-to-end path from
browser to container.

Add an HTTPS Listener on Port 443

The certificate is now in ACM, so the final task is to terminate TLS on the load balancer and replicate the host-based routing rules you created for HTTP.

1. **Open the ALB**

   Navigate to **EC2 ➤ Load Balancing ➤ Load Balancers**, choose **ecs-fargate-alb**, and switch to the **Listeners and rules** tab. You should see the existing **HTTP : 80** listener.

2. **Add the secure listener**

   Click **Add listener** (see Figure 10-26) and configure

   • **Protocol: HTTPS**

   • **Port:** Defaults to **443** (leave as is)

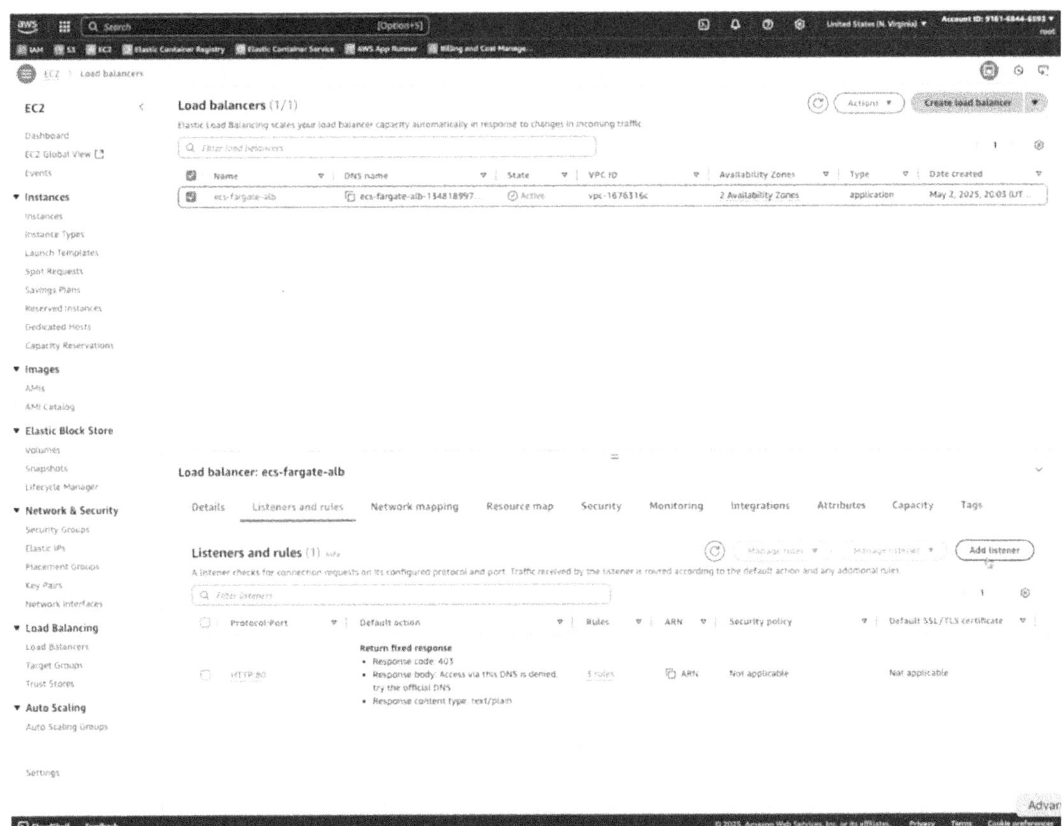

Figure 10-26. *Add listener*

3. **Default action: block ALB DNS access**

   Under **Routing actions**, pick **Return fixed response**.

   - **Response code: 403**

   - **Response body:** Enter a friendly note such as

   "Access via this DNS is denied. Please use the official domain."

   This denies traffic that reaches the ALB through its AWS-generated DNS name, maintaining the same restriction you applied on the HTTP listener.

4. **Attach the ACM certificate**

   Scroll to **Secure listener settings**.

   - **Certificate source:** Choose **From ACM**.

   - **Certificate:** Select the certificate you imported in this section; see Figure 10-27.

   Leave other TLS settings at their defaults and click ***Add***.

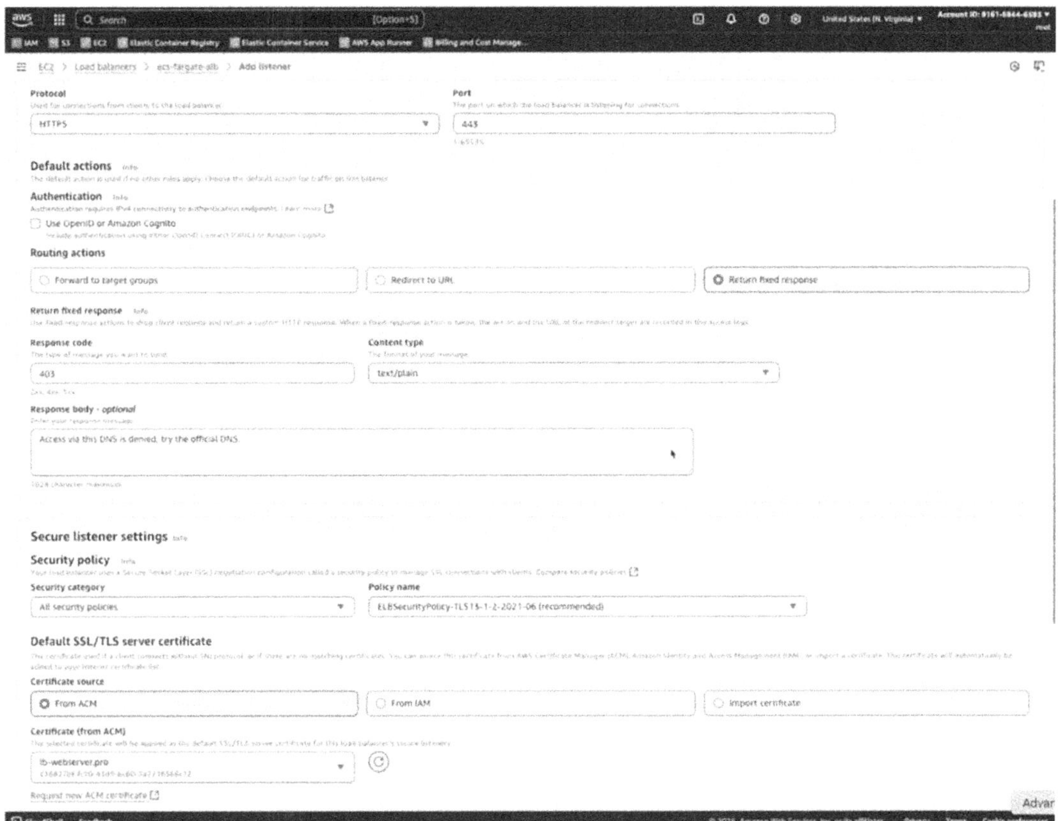

Figure 10-27. *Add a listener with imported ACM certificate*

5. **Recreate host-based rules for HTTPS**

   Using the knowledge from the section "Preparing the Ground for
   ECS (Elastic Container Service)," adjust the rules so that it's the
   same setup as before, except that now the listener is HTTPS : 443.
   The new listener starts with only the default rule you just defined.
   Add two more rules:

   - **Rule 1: Flask**

     - **Condition:** *Host header* equals `flask.lb-webserver.pro`

     - **Action:** *Forward to target group* → `flask-tg`

- **Rule 2: Streamlit**

    - **Condition:** *Host header* equals `streamlit.lb-webserver.pro`

    - **Action:** *Forward to target group* → `streamlit-tg`

6. **Verify the configuration**

    Your ***HTTPS : 443*** listener should now list three entries: the default 403 response and the two host header rules. The target groups remain `flask-tg` and `streamlit-tg`, identical to the HTTP setup; see Figure 10-28.

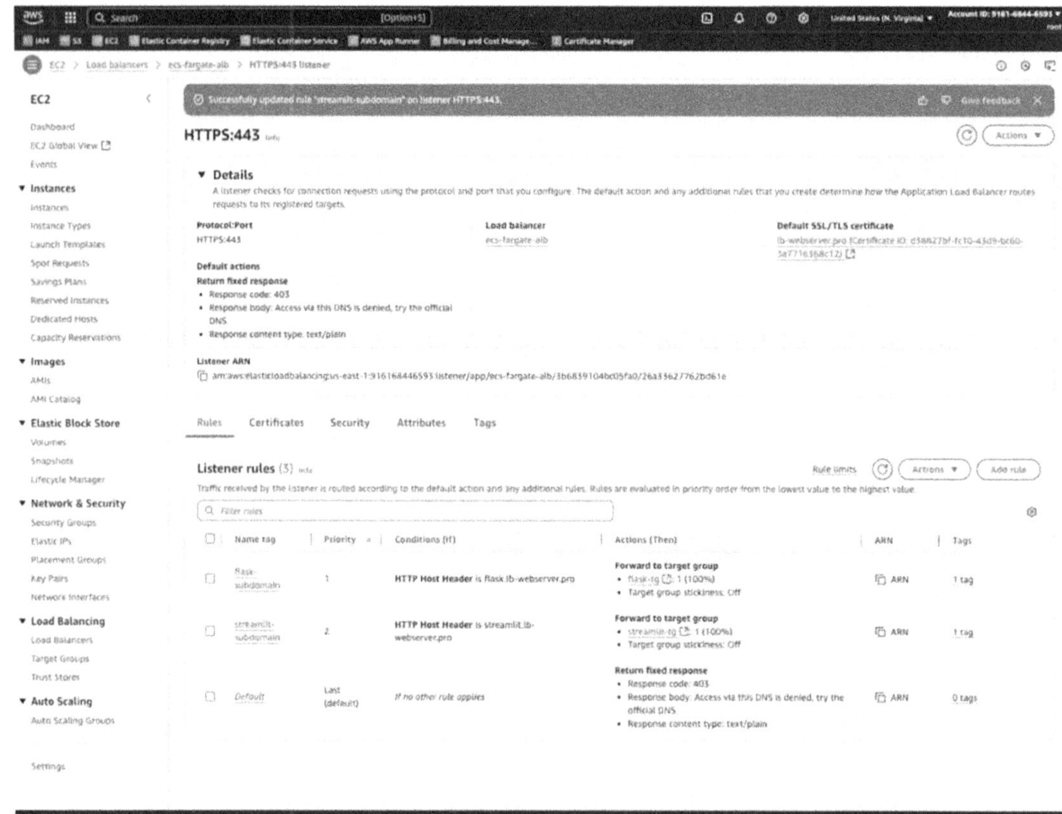

Figure 10-28. *HTTPS : 443 with rules*

At this point, both `http://` and `https://` requests for your subdomains follow the same routing logic, but only HTTPS traffic is encrypted. Because modern browsers automatically upgrade to HTTPS when certificates are present, you can optionally remove the HTTP listener.

Update the ALB Security Group for HTTPS Traffic

Until now the load balancer's security group (`alb-sg`) has allowed only HTTP on port 80. To serve encrypted traffic, you must switch that rule to HTTPS on port 443. No other security groups need modification because only the ALB is directly exposed to the internet.

1. In the EC2 console, open **Network & Security ➤ Security Groups** and select **alb-sg**.

2. Choose **Actions ➤ Edit inbound rules**.

3. Locate the existing rule that permits **HTTP / TCP / 80** from your IP range.

4. Change the **Type** field from **HTTP** to **HTTPS**. The **Port range** automatically updates from **80** to **443**. Leave the **Source** CIDR exactly as it is so you don't accidentally open the site to the whole world, unless that is your intention.

5. Click **Save rules**. The inbound rule list should now show **HTTPS / TCP / 443** with the same source CIDR; see Figure 10-29.

From this point forward, the ALB accepts only encrypted requests while preserving the IP restrictions you originally set.

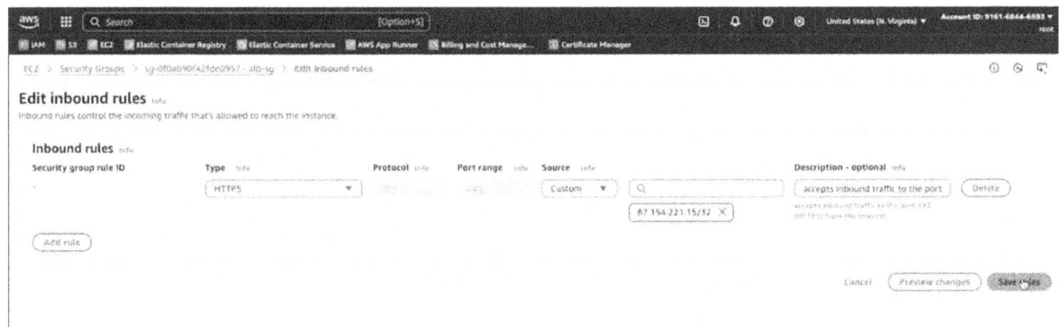

Figure 10-29. *Change the ALB's security group*

Switch the Services to the New HTTPS : 443 Listener

The final step is to point each ECS service at the secure listener you just created. Because the AWS console does not let you change a service's listener in place, you must **scale the service to zero tasks, delete it, and recreate it**, keeping the same target group but choosing the **HTTPS : 443** listener this time.

1) **Scale the existing services to zero and delete it**

   1. Go to **ECS ➤ Clusters ➤ ecs-fargate-cluster ➤ Services** and click **flask-app-service**.

   2. Choose **Update service**.

   3. In **Deployment configuration**, set **Desired tasks** to **0** (see Figure 10-30) and click **Update**.

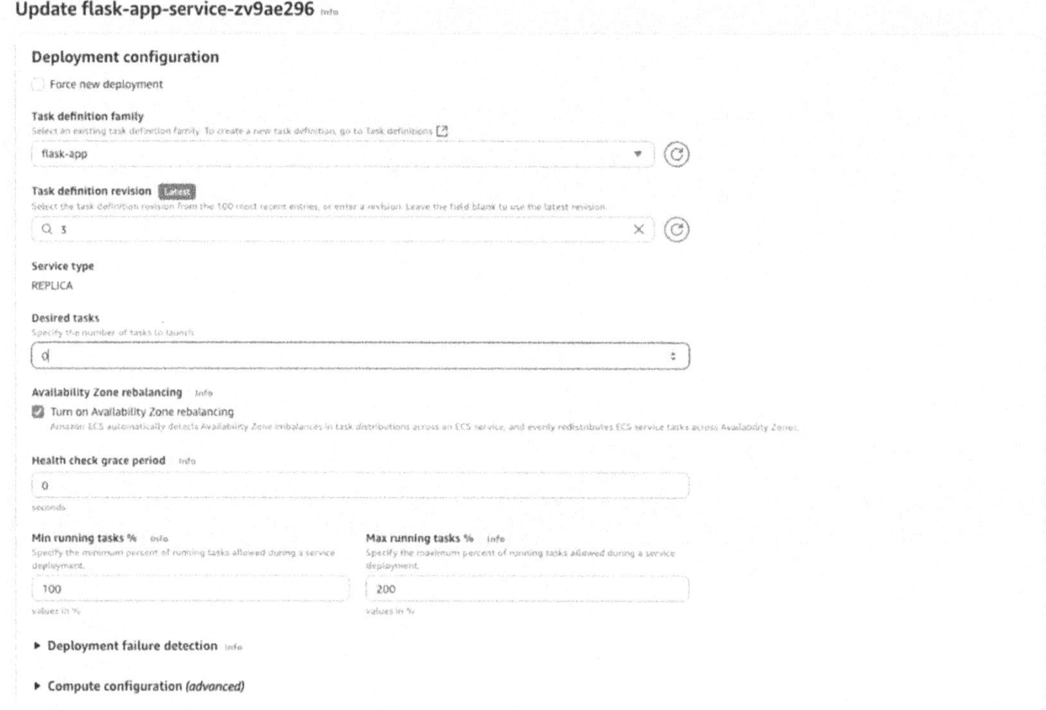

Figure 10-30. *Downscale the service's desired tasks to 0*

4. Wait until the task count shows **0 / 0**. Now that there are no tasks running for this service, you can delete it.

5. Back in the Services list (**Amazon Elastic Container Service ➤ Clusters ➤ ecs-fargate-cluster ➤ Services**), select the same service and choose **Delete**. Confirm the prompt.

6. Repeat the downscale-and-delete process for **streamlit-app-service**.

2) **Recreate the services with the HTTPS listener**

Use the same settings described in the section "Deployment with ECS and Fargate" with just two changes:

- In **Load balancing**, choose **HTTPS : 443** instead of the HTTP listener.

- Leave the **target group** exactly as before (flask-tg for Flask, streamlit-tg for Streamlit).

For clarity, here is the checklist for **Flask**; Streamlit is identical except for the task definition and target group. See Table 10-8.

Table 10-8. *Steps to recreating the service*

Wizard step	Key choices
Service details	*Task definition:* flask-app
Networking	*Security group:* ecs-sg-tasks (unchanged)
Load balancing	*Application Load Balancer:* ecs-fargate-alb *Listener:* HTTPS : 443 *Target group:* flask-tg
Create	Finish the wizard

Repeat for **Streamlit** (task definition streamlit-app, target group streamlit-tg).

3) **Verify HTTPS access**

When both new services show a green status, open a browser
and test:

- `https://flask.lb-webserver.pro`

- `https://streamlit.lb-webserver.pro`

You should receive the secure versions of each application, with the lock icon in the
address bar confirming TLS is active; see Figure 10-31. Your AWS serverless stack is now
fully operational and fully encrypted end to end.

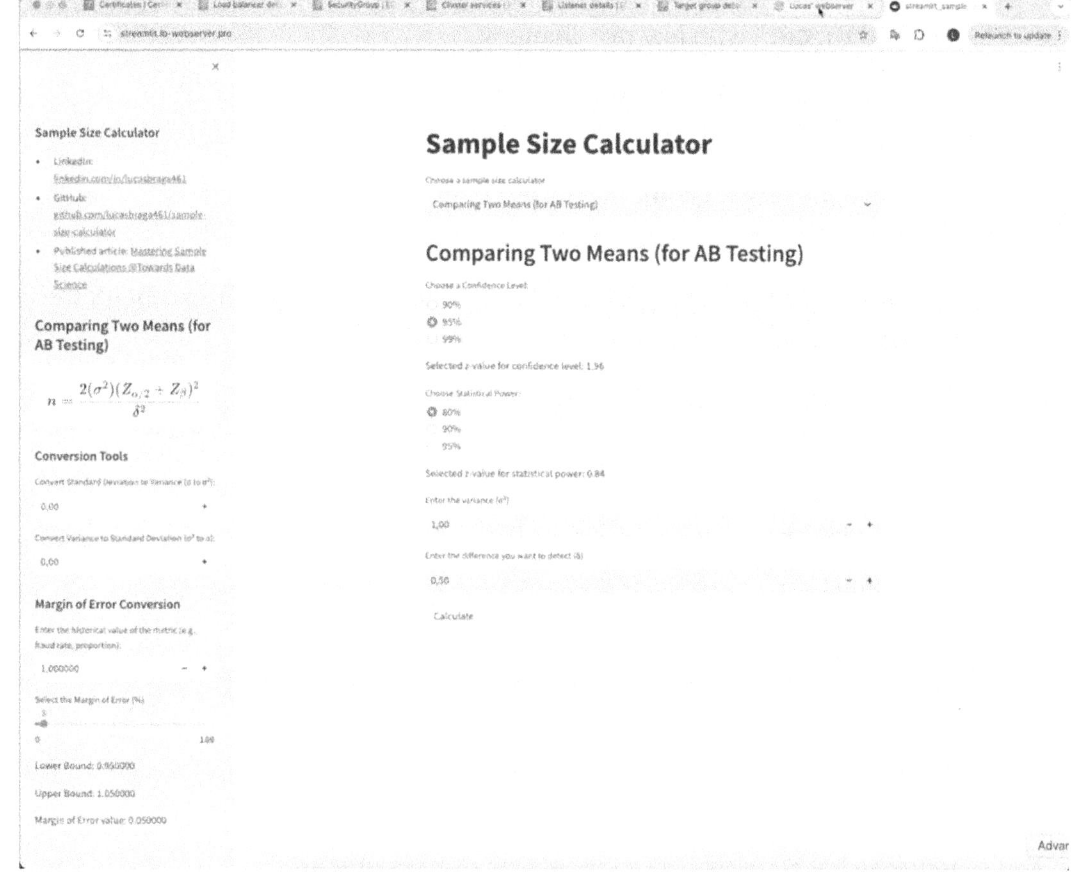

Figure 10-31. *Applications running with HTTPS*

Scaling Your Applications

Even in a serverless architecture you must decide how much compute you want each service to consume. In ECS Fargate, that capacity is expressed in two dimensions:

- **Task size**: The vCPU / memory pair you chose in the task definition (e.g., 2 vCPU + 4 GB for Flask)

- **Task count**: The number of identical tasks the service keeps running

You can raise throughput in two ways: **manually** by changing *desired tasks* or **automatically** with *service autoscaling*.

Manually Increasing the Desired Task Count

1. Open **ECS ➤ Clusters ➤ ecs-fargate-cluster ➤ Services** and click flask-app-service (or the Streamlit service).

2. Select **Update service**.

3. In **Deployment configuration**, set **Desired tasks** to the new value: 2, 3, 4, etc.

4. Scroll down and click **Update**.

Because each Flask task already reserves **2 vCPU** and **4 GB RAM**, raising *desired tasks* from 1 to 2 instantly doubles the compute available to that service (4 vCPU/8 GB total) and so on. The same logic applies to Streamlit, which consumes 1 vCPU/3 GB per task.

Manual scaling is quick and predictable, but in most production scenarios, you'll want capacity to grow and shrink automatically.

Configuring Service Autoscaling

Autoscaling adds or removes tasks based on a metric you choose, typically average CPU or memory usage.

1. Navigate to **ECS ➤ Clusters ➤ ecs-fargate-cluster ➤ Services**, select **flask-app-service**, and open the **Service auto scaling** tab.

2. Click **Set the number of tasks**; see Figure 10-32.

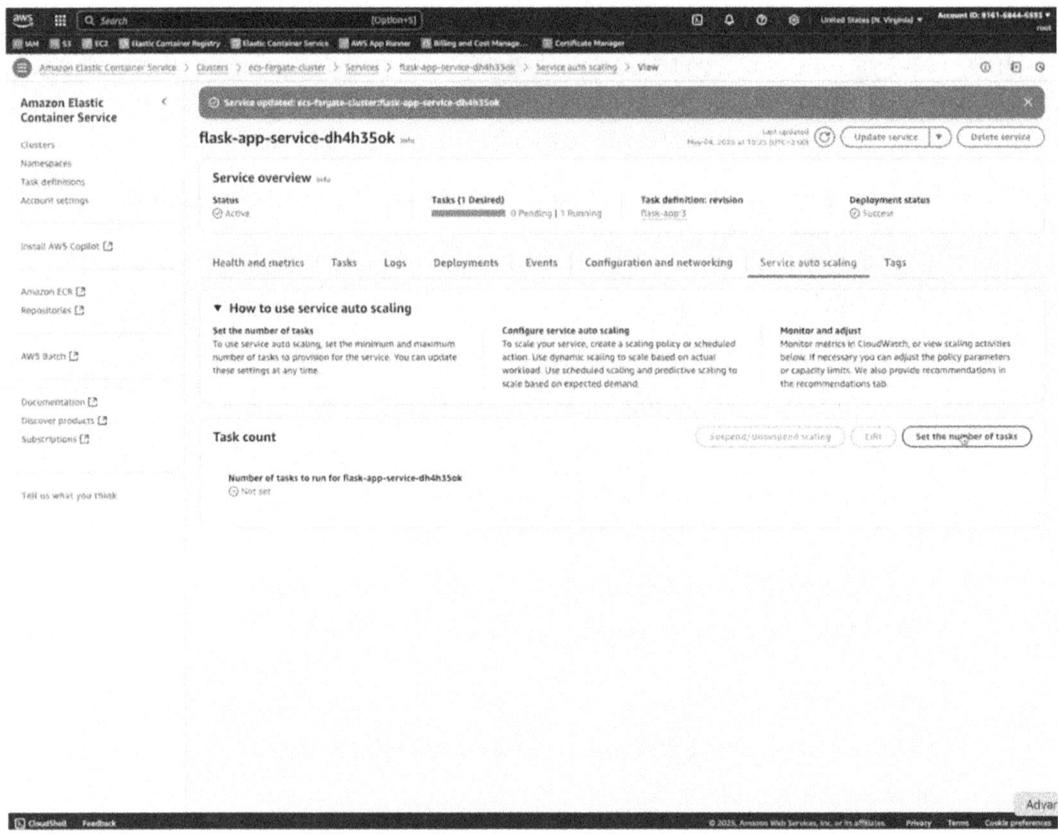

Figure 10-32. *Setting up autoscaling*

3. Tick **Use service auto scaling to adjust your service's desired task count**.

    - Set a ***minimum*** and ***maximum*** number of tasks (defaults are 1 and 10) and click ***Save***; see Figure 10-33.

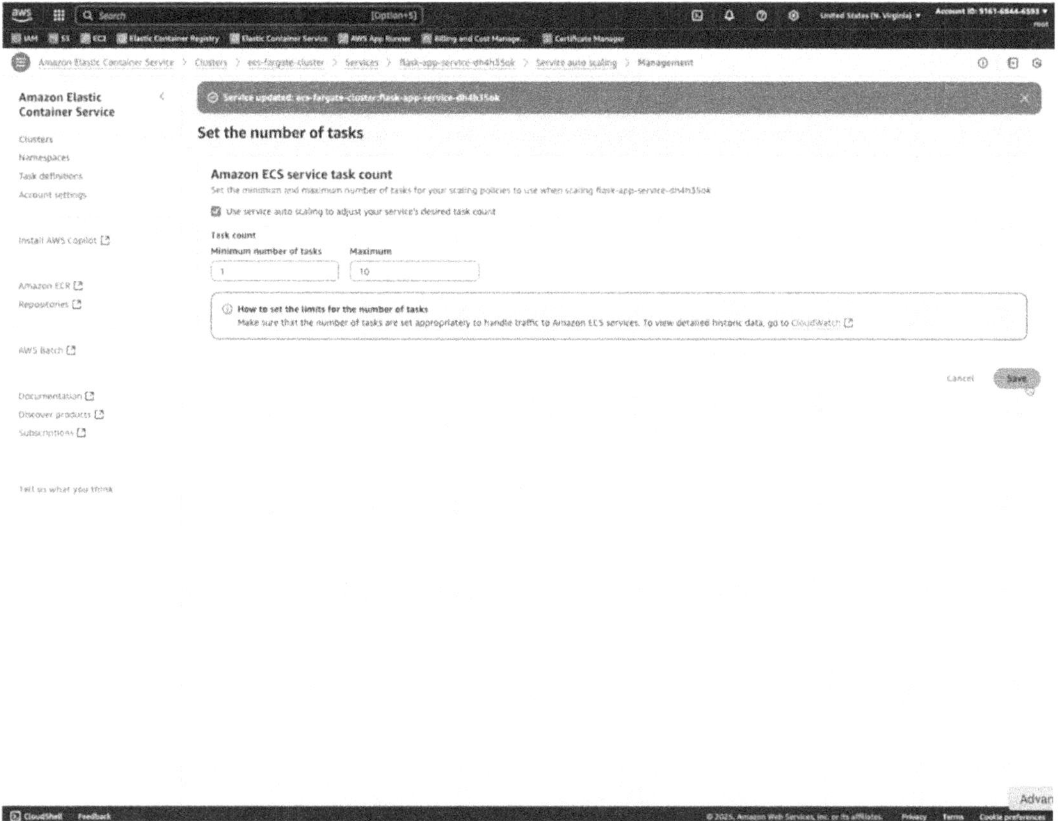

Figure 10-33. *Set minimum and maximum number of tasks*

4. Back on the autoscaling page, click **Create scaling policy**.

5. Leave **Policy type** at **Target tracking** and give the policy a name, for example, **alb-flask-auto-scaling-policy**.

6. **Metric type:** Pick **ECSServiceAverageCPUUtilization**.

7. **Target value:** Enter the utilization threshold, for example, **50 percent**.

    Meaning: when average CPU reaches 50 %, ECS adds tasks; when it falls well below, ECS removes them.

8. Click ***Create scaling policy***; see Figure 10-34.

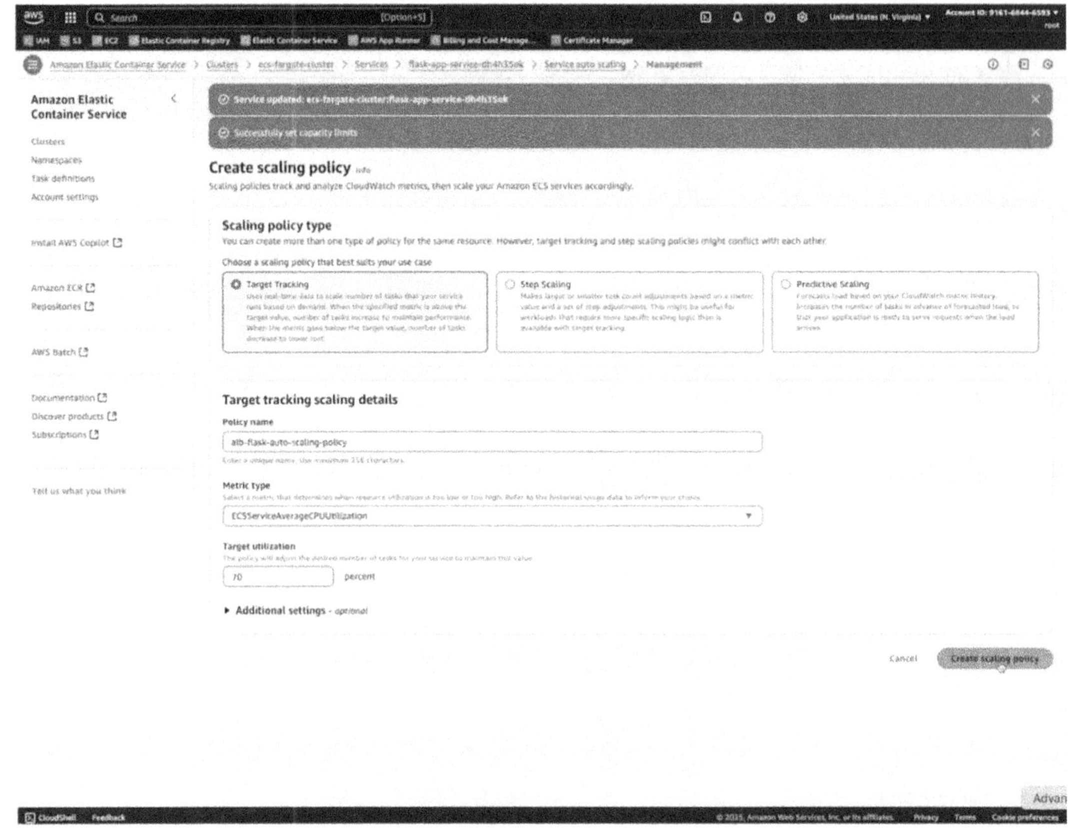

Figure 10-34. *Choose a target utilization*

Repeat the same steps for the Streamlit service if desired. From now on, each service adjusts its task count dynamically within the bounds you set, keeping response times snappy without wasting compute on quiet days.

Your serverless deployment now scales up for traffic spikes and backs down to reduce cost, completing the full production-ready setup.

Summary

In this chapter, you deployed Flask and Streamlit on **AWS ECS Fargate** and secured them using HTTPS via an **Application Load Balancer (ALB)**. Specifically, you

- Created an IAM user and access keys with permission to manage ECS, ECR, and related services

- Installed and configured the AWS CLI locally with a custom profile (`serverless-user`)

    - AWS CLI Installation documentation: `https://docs.aws.amazon.com/cli/latest/userguide/install-cliv2.html`

- Created **ECR repositories** for both Flask and Streamlit, then built, tagged, and pushed the Docker images using the AWS CLI

    - Amazon ECR User Guide documentation: `https://docs.aws.amazon.com/AmazonECR/latest/userguide/what-is-ecr.html`

- Created the required **security groups**, one for the ALB (`alb-sg`) and another for ECS containers (`ecs-sg-tasks`)

    - Application Load Balancers documentation: `https://docs.aws.amazon.com/elasticloadbalancing/latest/application/introduction.html`

- Set up **target groups** for both applications and configured a new **Application Load Balancer (ALB)** with host-based routing rules

- Denied access to the default AWS DNS by returning HTTP 403 responses

- Configured **host-based routing** for `flask.lb-webserver.pro` and `streamlit.lb-webserver.pro` to forward requests to their respective target groups

- Created ECS **task definitions** and launched both services using **Fargate**, attaching them to the correct target groups

    - AWS ECS Fargate documentation: `https://docs.aws.amazon.com/AmazonECS/latest/userguide/what-is-fargate.html`

- Updated **Namecheap DNS** settings to use CNAME records pointing to the ALB DNS name

- Imported an **SSL certificate** from Namecheap into **AWS Certificate Manager (ACM)**

- AWS Certificate Manager (ACM) documentation: `https://docs.aws.amazon.com/acm/latest/userguide/gs-acm-import.html`

- Added an **HTTPS listener** to the ALB and mirrored host-based routing logic to secure the services

- Updated the ALB's **security group** to allow HTTPS (port 443) and removed HTTP access

- Redeployed both ECS services to connect to the new HTTPS listener

- Verified access to both services via `https://flask.lb-webserver.pro` and `https://streamlit.lb-webserver.pro`

- Configured **manual scaling** by adjusting the desired task count

  - ECS Auto Scaling documentation: `https://docs.aws.amazon.com/AmazonECS/latest/developerguide/service-auto-scaling.html`

- Enabled **autoscaling policies** based on CPU utilization using ECS target tracking

In the next three chapters (11–13), you'll move from deployment into **real-world demos**. In Chapter 11, you'll explore Jenkins as an ETL/ELT platform for data science workflows. In Chapter 12, you'll use Streamlit as a front-end dashboard to visualize machine learning results. And in Chapter 13, you'll use Flask to expose a trained ML model as an API endpoint, demonstrating how each component fits into a practical MLOps pipeline.

## PART III

Jenkins, Streamlit, and Flask Demos

Demo: Using Jenkins As an ETL/ELT Platform for Data Science

In Chapters 9 and 10, you explored how to deploy your Flask and Streamlit applications to serverless platforms, first using **Google Cloud Run** and then with **AWS ECS and Fargate**. Both chapters walked through containerization, HTTPS configuration, load balancing, and secure DNS-based routing of production-ready services.

In this chapter, you will implement a hands-free ETL/ELT pipeline using Jenkins and Python. You will understand what ETL and ELT are; you will create AWS and GCP credentials for Jenkins to read and write cloud storage buckets. You will configure buckets and upload files to simulate daily data ingestion. You will write a Python-based ETL pipeline that runs in Jenkins, and you will validate the results confirming the output CSV files landed correctly in the /output/ folders on AWS S3 and GCP buckets.

Data science projects often begin with an **ETL or ELT pipeline**, the mechanism that makes raw data usable. Imagine you are a data scientist at a fintech company. Customer payment transactions flow through several external payment gateways; those gateways collect the raw transaction records and agree to hand the data back to your team for analytics. How that hand-off happens varies:

- **Daily CSV drops:** A third-party company writes one file per day into an **AWS S3** bucket or a **Google Cloud Storage** bucket you control.

- **Pull-style API:** A more robust approach, a third-party company exposes a REST endpoint that returns transactions in JSON when you call it.

© Lucas H. Benevides e Braga 2025
L.H.B.e. Braga, *Deploying Secure Data Science Applications in the Cloud*,
https://doi.org/10.1007/979-8-8688-1715-1_11

Either way the data ends up "somewhere out there," and you need to turn it into insight. In many workflows, you will

1. **Extract** raw files or API payloads from the source location

2. **Transform** the data: cleaning, deduplicating, reformatting, aggregating, or joining with reference tables

3. **Load** the polished dataset to a destination folder, database, or data lake for analysis

We start by setting up the necessary cloud credentials and storage locations in the next two sections, "Cloud Credentials for the ETL Pipeline" and "Create Buckets and Upload Test Files"; then in the subsequent section, "ETL Python Code Walkthrough," we'll examine the Python code and deploy the ETL job in Jenkins.

Note   ETL vs. ELT: What's the Difference?

- **ETL: Extract ➤ Transform ➤ Load**

   You clean and re-shape the data *before* it touches your production storage, saving space at the cost of flexibility.

- **ELT: Extract ➤ Load ➤ Transform**

   You dump the raw data directly into cheap, scalable storage (S3, GCS, BigQuery, Snowflake, etc.) and transform it later on demand.

The choice hinges on infrastructure:

- If your organization has abundant, low-cost object storage and powerful cloud databases, **ELT** offers maximum agility; you can revisit raw data whenever the business asks new questions.

- If storage is tight or expensive, **ETL** may be safer. Transform first, keep only what you need, and avoid cluttering disks with uncurated files.

Throughout the remainder of this chapter, you will see both patterns and understand when each makes sense in practice.

Cloud Credentials for the ETL Pipeline

Before Jenkins can pull data from cloud storage, it needs credentials for both AWS and Google Cloud. You generated AWS keys in Chapter 10 (sections "Create an IAM User and Access Keys" and "Install and Configure the AWS CLI"); we will reuse those and then create an equivalent service-account key in GCP.

Copy Your AWS Profile into the Project

The profile you created earlier already has **AmazonS3FullAccess**, which is enough to read and write objects for this demo.

1. Locate ~/.aws/config on your workstation.

2. Copy the file into the repository under **deploy-secure-ds-apps-book/jenkins/.aws/**.

3. (Optional) Rename the profile inside the file from serverless-user to something clearer, for example, book-etl.

   Code-block 11-1 shows the structure of your working directory.

Code-block 11-1. Folder structure with the .aws/config file

```
├ deploy-secure-ds-apps-book
├── jenkins
│       └── .aws
│             └── config
...
```

Create a GCP Service-Account Key

Automated jobs need non-interactive credentials. A **service account** supplies those.

1. Open **IAM & Admin ➤ Service accounts** in the Google Cloud console.

2. Click **Create Service Account** and fill in; see Table 11-1.

Table 11-1. *Fill in to create a service account*

Field	Example value
Service account name	`etl-service-account`
Description	*Service account for ETL bucket access*

3. Click **Create and continue**.

4. Under **Permissions**, expand **Cloud Storage** and add **one** of the following:

   - **Storage Object Viewer:** Read-only access.

   - **Storage Object Admin:** Read, write, and delete.

     Follow the compliance principle of least privilege and choose the narrowest scope that meets your needs.

5. Also add **Storage Bucket Viewer** so the account can list buckets; see Figure 11-1.

6. Click **Continue ➤ Done** to finish.

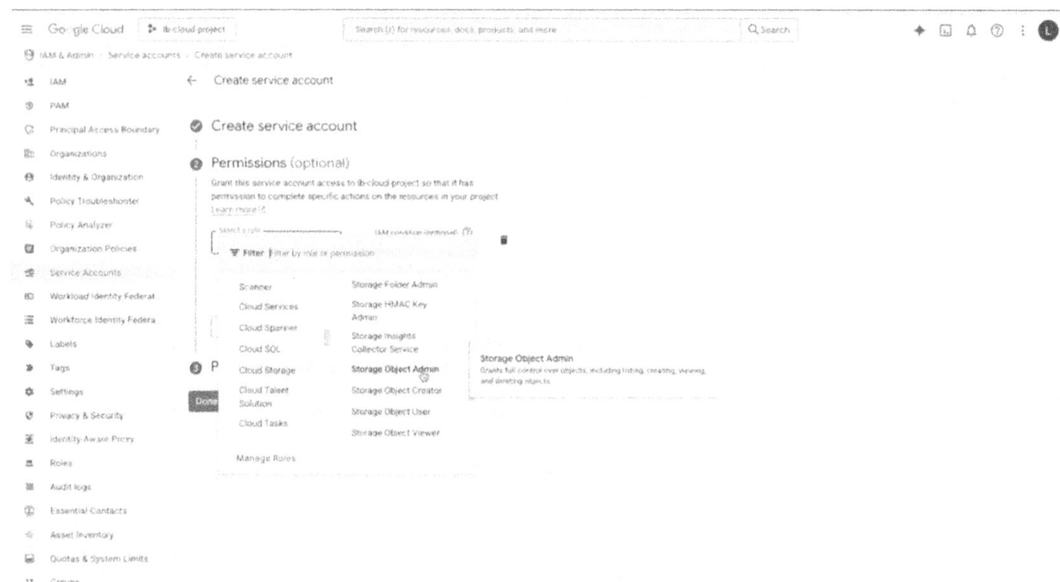

Figure 11-1. *Choose the Service Account's Permission*

Generate and Store the JSON Key

1. Still on the Service accounts page, click the new account.

2. Open the **Keys** tab ➤ **Add key** ➤ **Create new key**.

3. Select **JSON** and click **Create**. The file downloads automatically–store it someplace safe.

Add the key to the repository under jenkins/config/; refer to Code-block 11-2.

Code-block 11-2. Folder structure with .env and config/

```
⊢ deploy-secure-ds-apps-book
├── jenkins
│   ├── .aws
│   │   └── config
│   ├── config
│   │   └── 1b-cloud-project-65885ec00848.json
```

Important   Add the JSON file name to .gitignore so you don't publish credentials to your remote repo.

With both the AWS profile and the GCP service-account key in place, the Jenkins Docker image you build in the next section will have everything it needs to authenticate to either cloud and run the ETL tasks on schedule.

Create Buckets and Upload Test Files

Before Jenkins can run its first pipeline, we need somewhere to land the raw data. In the real world, your payments partner would push files into a bucket you provide; for the demo, we will create the buckets ourselves and upload a handful of CSVs to mimic a day's drop.

The AWS Side

Open the S3 console, check the region selector in the top right, and **create a bucket** called something like **book-project**. The wizard's defaults–private access, no versioning, and standard storage class–are fine, so scroll straight to the bottom and click **Create bucket**.

With the bucket in place, click its name, choose **Upload**, and add two or three small CSV files. They are only placeholders, but the presence of real objects will make it obvious when the ETL job has done its work.

In production, your partner needs a way to deliver fresh files every day. A common pattern is to create a dedicated IAM user, grant it a narrowly scoped policy (e.g., `s3:PutObject` and `s3:ListBucket` on `arn:aws:s3:::book-project/*`), and hand the resulting access keys to the partner. They can then automate uploads with the AWS CLI, SDK, or a simple `curl` against a pre-signed URL.

The Google Cloud Side

Now switch hats to Google Cloud Storage. Open **Storage ➤ Buckets**, confirm the correct GCP project is selected, and click **Create**. Give the new bucket a name, `book-etl-project` works, and accept the following default choices:

- **Location type:** *Multi-region* (a click on **Continue** advances the wizard)

- **Storage class:** *Standard*

All other fields can stay untouched; click **Create**.

Inside the bucket, press **Upload** and drop in the same CSV files you used on S3. The object sizes are irrelevant; we just want Jenkins to find something waiting for it.

When a third-party vendor needs to populate this bucket in real life, you would create a **service account**, as we did in this chapter, section "Cloud Credentials for the ETL Pipeline," and give it the role **Storage Object Creator** (write-only) or **Storage Object Admin** (full control) on this bucket. The vendor authenticates with the JSON key you hand over, and your security posture remains tight because that key is good for exactly one job: writing objects to this bucket.

With both buckets stocked with sample data, we can turn our attention to the Python code that will pull, transform, and reload those files and then wrap the whole flow in Jenkins so it runs every morning without human intervention.

Note Choosing a Google Cloud Storage Class   Google's four storage classes balance cost against how quickly and how often you plan to read the data; see Table 11-2.

Table 11-2. *GCP's different storage classes*

Class	Typical access pattern	Use-case examples	Cost and performance
Standard	Multiple reads per week or day	Daily ETL feeds, real-time dashboards	Fastest retrieval, highest price per GB
Nearline	A few reads per **month**	Monthly backups, point-in-time snapshots	Cheaper than Standard, small access fee per read
Coldline	One read per **quarter**	Disaster-recovery copies, quarterly compliance exports	Low storage cost, higher retrieval fee and latency
Archive	One read per **year** (or rarely ever)	E-discovery evidence, long-term audit logs	Lowest storage cost, highest retrieval fee, 365-day minimum retention per object

---

Can you touch Archive data more often?   Technically yes, Google won't block you, but every read incurs a hefty access charge and defeats the purpose of the tier. If you expect to reopen the files more than a couple of times a year, Coldline or Nearline almost always delivers a better cost-to-convenience ratio.

Choose the class that matches the *real* cadence of your workloads; you'll save money without slowing down your analysts.

ETL Python Code Walkthrough

Before we open the script itself, Jenkins needs two new bind-mounts so it can "see" your configuration files and the ETL source code at runtime. Open **docker-compose.yaml** and add these two lines under volume (volumes is located within jenkins within services), see Code-block 11-3.

Code-block 11-3. Added lines to the docker-compose.yaml file

```
# docker-compose.yaml  (excerpt)
services:
  jenkins:
    ...
    volumes:
      - ./jenkins/config:/app/config
      - ./jenkins/etl_scripts:/app/etl_scripts
```

These lines map

- **./jenkins/config/** -> **/app/config/** inside the container, where the GCP JSON key will live

- **./jenkins/etl_scripts/** -> **/app/etl_scripts/**, giving the container access to etl_job.py and any helper modules

With the mount points in place, confirm that every Python dependency required by etl_job.py is listed in **jenkins/requirements.txt**. If you later add a library, say, pandas or boto3, append it to that file and rebuild the Jenkins image; see Code-block 11-4.

Code-block 11-4. Commands to install more libraries for Jenkins to operate

```
docker-compose build jenkins
docker-compose up -d
```

The GitHub repo now includes **jenkins/etl_scripts/etl_job.py**, which orchestrates the full workflow:

1. **Extract:** Downloads the raw CSV from either S3 or Cloud Storage, depending on a runtime flag

2. **Transform:** Cleans the data, applies schema changes, aggregates totals, removes duplicates, whatever your business logic requires

3. **Load:** Writes the polished file back to a /processed/ folder in the same bucket

In the next part, we'll dissect etl_job.py section by section so you can see exactly how each stage works.

Inside etl_job.py—Imports, Configuration, and I/O Helpers

The script begins with a tight set of imports, pandas for data wrangling, boto3 for AWS calls, and google.cloud.storage for GCP. Immediately after, it declares a few constants that everything else relies on (Code-block 11-5).

Code-block 11-5. etl_job.py initial part

```
import pandas as pd
import boto3
from google.cloud import storage
from datetime import datetime, timedelta
from io import BytesIO
from pathlib import Path
import argparse

# === Config ===
CREDENTIALS_PATH = Path('../config/lb-cloud-project-65885ec00848.json')
AWS_BUCKET_NAME = 'book-project-54641'
GCP_BUCKET_NAME = 'book-etl-project'
```

- **CREDENTIALS_PATH** points to the JSON key you generated in this chapter, section "Cloud Credentials for the ETL Pipeline."

- **AWS_BUCKET_NAME** and **GCP_BUCKET_NAME** match the buckets you created in the section "Create Buckets and Upload Test Files."

Extraction Helpers

Two functions pull the raw CSV: one from S3 and one from Cloud Storage (Code-block 11-6).

Key points

- The S3 call uses the **book-etl** profile you added to ~/.aws/config.

- The GCP call instantiates a client from the service-account key that keeps credentials outside the code.

Code-block 11-6. etl_job.py extraction functions

```
# === Extraction Functions ===

def extract_from_s3(bucket_name, file_name):
    print(f"Extracting {file_name} from AWS S3 bucket {bucket_name}")
    s3 = boto3.Session(profile_name='book-etl').client('s3')
    response = s3.get_object(Bucket=bucket_name, Key=file_name)
    data = response['Body'].read()
    return pd.read_csv(BytesIO(data))

def extract_from_gcp(bucket_name, file_name, credentials_path):
    print(f"Extracting {file_name} from GCP bucket {bucket_name}")
    client = storage.Client.from_service_account_json(credentials_path)
    bucket = client.get_bucket(bucket_name)
    blob = bucket.blob(file_name)
    data = blob.download_as_bytes()
    return pd.read_csv(BytesIO(data))
```

Upload Helpers

The companion functions push the transformed DataFrame back to each cloud; see Code-block 11-7. They appear *before* the transform logic only because they're structurally similar to the extractors; the main routine still calls them last (E → T → L).

Code-block 11-7. etl_job.py upload functions

```python
# === Upload Functions ===

def upload_to_s3(df, bucket_name, file_name):
    print(f"Uploading {file_name} to AWS S3 bucket {bucket_name}")
    buffer = BytesIO()
    df.to_csv(buffer, index=False)
    buffer.seek(0)
    s3 = boto3.Session(profile_name='book-etl').client('s3')
    s3.put_object(Bucket=bucket_name, Key=file_name, Body=buffer.getvalue())
    print("Upload to S3 completed.")

def upload_to_gcp(df, bucket_name, file_name, credentials_path):
    print(f"Uploading {file_name} to GCP bucket {bucket_name}")
    buffer = BytesIO()
    df.to_csv(buffer, index=False)
    buffer.seek(0)
    client = storage.Client.from_service_account_json(credentials_path)
    bucket = client.get_bucket(bucket_name)
    blob = bucket.blob(file_name)
    blob.upload_from_file(buffer, content_type='text/csv')
    print("Upload to GCP completed.")
```

Both functions serialize the DataFrame in memory and stream it to cloud storage, no intermediate disk writes, so this build step remains stateless and portable. (The Jenkins controller itself is still stateful, which is why we run it on a VM with persistent storage.)

In the next part, we'll look at the transformation step and the command-line interface that glues these helpers into a single end-to-end pipeline.

What the Raw Data Looks like

Before you can appreciate the transformation logic, it helps to see a slice of the input file; see Figure 11-2. Each CSV contains seven columns; see Table 11-3.

	transaction_id	user_id	amount	currency	date	status	payment_gateway
0	81	1081	60.00	USD	2025-05-17	completed	PayPal
1	82	1082	110.50	EUR	2025-05-17	completed	Stripe
2	83	1083	40.75	USD	2025-05-17	failed	Payoneer
3	84	1084	90.00	USD	2025-05-17	completed	Stripe
4	85	1085	47.00	EUR	2025-05-17	completed	PayPal
5	86	1086	25.50	USD	2025-05-17	completed	Stripe
6	87	1087	80.80	EUR	2025-05-17	completed	Payoneer
7	88	1088	100.00	USD	2025-05-17	completed	PayPal
8	89	1089	56.60	EUR	2025-05-17	completed	Stripe
9	90	1090	75.75	USD	2025-05-17	completed	Payoneer
10	91	1091	NaN	EUR	2025-05-17	completed	Stripe
11	92	1092	65.00	USD	2025-05-17	completed	PayPal
12	93	1093	45.45	EUR	2025-05-17	completed	Payoneer
13	94	1094	55.55	USD	2025-05-17	completed	Stripe
14	95	1095	70.70	EUR	2025-05-17	completed	PayPal
15	96	1096	85.85	USD	2025-05-17	completed	Stripe
16	97	1097	95.00	EUR	2025-05-17	failed	Payoneer
17	98	1098	30.30	USD	2025-05-17	completed	PayPal
18	99	1099	120.00	EUR	2025-05-17	completed	Stripe
19	100	1100	40.00	USD	2025-05-17	completed	Payoneer

Figure 11-2. *First 20 rows of the raw CSV transactions file*

Table 11-3. *Raw data file explained*

Column	Type	Meaning
transaction_id	**int**	Primary key—unique per payment
user_id	**int**	Unique identifier for the customer
Amount	**float**	Transaction amount in the original currency
Currency	**str**	ISO currency code, for example, *USD*, *EUR*
Date	**str**	Day of the transaction, YYYY-MM-DD
Status	**str**	*completed* or *failed*
payment_gateway	**str**	Provider that processed the payment—PayPal, Stripe, Payoneer, etc.

The Transformation Pipeline

transform_data(), from Code-block 11-8, cleans the file and derives two daily KPIs:

1. **Success rate (counts):** How many payments completed vs. attempted

2. **Success rate (amount):** How many dollars (USD adjusted) cleared vs. attempted

Key steps in the function

- **De-duplicate** rows on the tuple (user_id, amount, currency, date, status, payment_gateway).

- **Drop null amounts:** Guard against malformed rows.

- **Normalize currency:** Convert EUR to USD with a static rate map.

- **Group by date + payment_gateway** to compute

  - Total transactions

  - Successful transactions

  - Success-rate %

  - Total amount (USD)

  - Successful amount (USD)

  - Amount-success-rate %

It returns **two DataFrames**: one for counts and one for dollar amounts.

Code-block 11-8. etl_job.py transform functions

```
# === Transformation Function ===

def transform_data(df):
    print("Transforming data...")

    df.drop_duplicates(subset=['user_id', 'amount', 'currency', 'date',
    'status', 'payment_gateway'], inplace=True)
    df.dropna(subset=['amount'], inplace=True)
```

```
conversion_rates = {'EUR': 1.1, 'USD': 1.0}
df['amount_usd'] = df.apply(lambda x: round(x['amount'] *
conversion_rates[x['currency']], 2), axis=1)

# Success count rate
success_counts = df[df['status'] == 'completed'].groupby(['date',
'payment_gateway']).size()
total_counts = df.groupby(['date', 'payment_gateway']).size()

success_rate_df = pd.DataFrame({
    'date': total_counts.index.get_level_values('date'),
    'payment_gateway': total_counts.index.get_level_values
    ('payment_gateway'),
    'total_transactions': total_counts.values,
    'successful_transactions': success_counts.reindex
    (total_counts.index, fill_value=0).values
})
success_rate_df['success_rate_percent'] = round(
    (success_rate_df['successful_transactions'] /
    success_rate_df['total_transactions']) * 100, 2
)

# Success amount rate
total_amount = df.groupby(['date', 'payment_gateway'])
['amount_usd'].sum()
successful_amount = df[df['status'] == 'completed'].groupby
(['date', 'payment_gateway'])['amount_usd'].sum()

amount_success_rate_df = pd.DataFrame({
    'date': total_amount.index.get_level_values('date'),
    'payment_gateway': total_amount.index.get_level_values
    ('payment_gateway'),
    'total_amount_usd': total_amount.values,
    'successful_amount_usd': successful_amount.reindex
    (total_amount.index, fill_value=0).values
})
amount_success_rate_df['amount_success_rate_percent'] = round(
```

```
        (amount_success_rate_df['successful_amount_usd'] /
        amount_success_rate_df['total_amount_usd']) * 100, 2
    )

    return success_rate_df, amount_success_rate_df
```

The Main Driver: run_etl()

The function run_etl() (see Code-block 11-9) stitches everything together. Table 11-4 puts together a summary of the ETL process.

Table 11-4. *Summary of the ETL process*

Phase	What it does
Extract	Builds a list of the last three day's file names (raw_transactions_YYYY_MM_DD.csv). Uses either extract_from_s3() or extract_from_gcp() depending on the --source flag
Transform	Passes the concatenated DataFrame to transform_data(), receiving the two KPI tables
Load	Writes each KPI table to an /output/ folder in the same bucket (upload_to_s3() or upload_to_gcp())

Code-block 11-9. etl_job.py run_etl function

```python
# === Main ETL Logic ===

def run_etl(source, today, credentials_path):
    ### 1. Extract (E) data
    today_minus_1 = (today - timedelta(days=1)).strftime('%Y_%m_%d')
    date_list = [(today - timedelta(days=i)).strftime('%Y_%m_%d') for i in
    range(1, 4)]
    file_names = [f"raw_transactions_{date}.csv" for date in date_list]

    df_ls = []
    for file in file_names:
        if source == 'aws':
            df_ls.append(extract_from_s3(AWS_BUCKET_NAME, file))
```

```
    elif source == 'gcp':
        df_ls.append(extract_from_gcp(GCP_BUCKET_NAME, file,
        credentials_path))

df = pd.concat(df_ls, ignore_index=True)

### 2. Transform (T) data
success_rate_df, amount_success_rate_df = transform_data(df)

### 3. Load (L) transformed data to bucket
folder = 'output'
if source == 'aws':
    upload_to_s3(success_rate_df, AWS_BUCKET_NAME,
    f"{folder}/success_rate_df_{today_minus_1}.csv")
    upload_to_s3(amount_success_rate_df, AWS_BUCKET_NAME,
    f"{folder}/amount_success_rate_df_{today_minus_1}.csv")
elif source == 'gcp':
    upload_to_gcp(success_rate_df, GCP_BUCKET_NAME,
    f"{folder}/success_rate_df_{today_minus_1}.csv", credentials_path)
    upload_to_gcp(amount_success_rate_df, GCP_BUCKET_NAME,
    f"{folder}/amount_success_rate_df_{today_minus_1}.csv",
    credentials_path)
```

The variable today_minus_1 lets the script name the output files with yesterday's date, matching the raw input it just processed.

Command-Line Entry Point

parse_args() exposes two optional flags; see Code-block 11-10.

- --source aws | gcp # default gcp
- --date YYYY-MM-DD # default = today; handy for back-fills

Code-block 11-10.   etl_job.py final functions

```
# === CLI Parser ===

def parse_args():
    parser = argparse.ArgumentParser(description="ETL pipeline to extract,
    transform, and load transaction data.")
```

```
parser.add_argument("--source", choices=["aws", "gcp"], default="gcp",
help="Cloud source. Default is gcp.")
parser.add_argument("--date", type=str, help="Base date in YYYY-MM-DD
format. Defaults to today.")
return parser.parse_args()

# === Entry Point ===

if __name__ == "__main__":
    args = parse_args()

    source = args.source
    if args.date:
        today = datetime.strptime(args.date, "%Y-%m-%d")
    else:
        today = datetime.today()

    run_etl(source, today, CREDENTIALS_PATH)
```

Python Dependencies

Requirements.txt (see Code-block 11-11) contains exactly the libraries the script imports.

Code-block 11-11. requirements.txt

```
pandas==2.0.3
numpy==1.24.4
boto3==1.34.94
google-cloud-storage==2.16.0
```

If you ever add another import to etl_job.py, add the package here and rebuild Jenkins.

Recap of the Jenkins Folder Layout

Your jenkins/ directory should now look like this; see Code-block 11-12.

Code-block 11-12. deploy-secure-ds-apps-book/jenkins folder structure

```
├── .aws
│    └── config
├── Dockerfile
├── config
│    └── lb-cloud-project-65885ec00848.json
├── etl_scripts
│    └── etl_job.py
├── requirements.txt
```

Deploying the ETL Job in Jenkins

With the script and credentials in place, the final step is to wire the pipeline into Jenkins so it runs on a schedule, and whenever you need a back-fill.

Create a New Job

Open your Jenkins dashboard (e.g., `https://jenkins.lb-webserver.pro`) and click + **New Item**; see Figure 11-3.

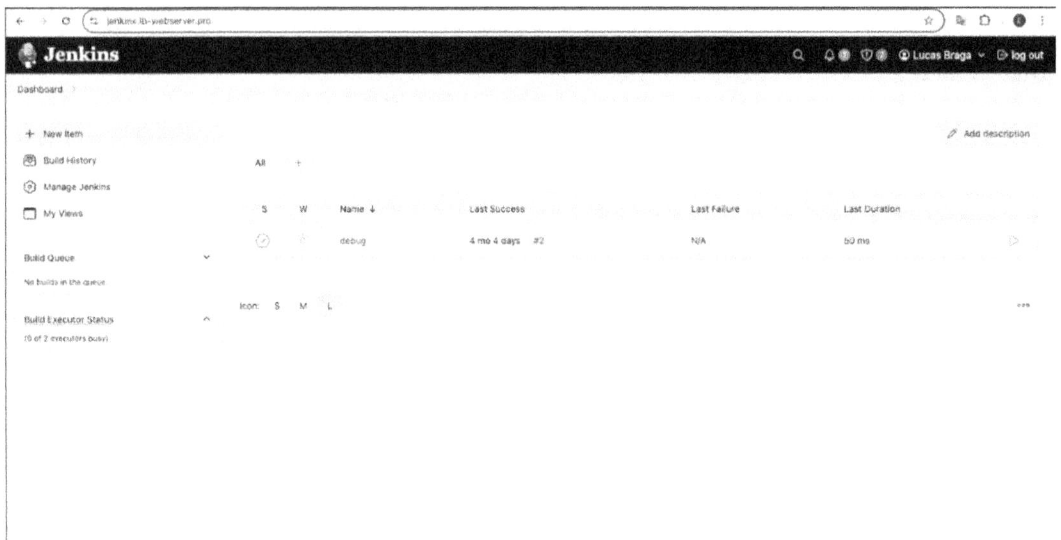

Figure 11-3. *Dashboard in Jenkins UI*

- **Name** the job **payment-transactions-etl**.

- Select **Freestyle project**.

- Click **OK**; see Figure 11-4.

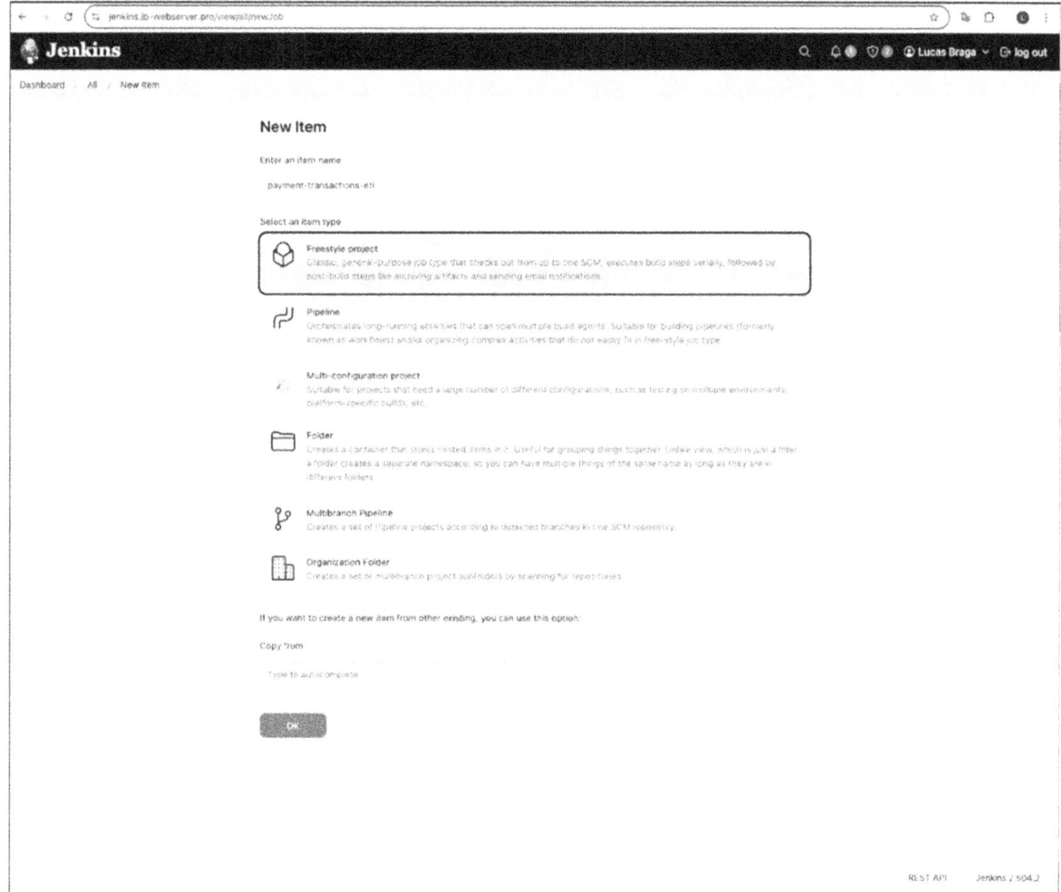

Figure 11-4. *Create a new ETL job.*

Add a short description such as

- *Daily ETL job that extracts raw CSVs from S3 or GCS, transforms them, and uploads the results to /output/.*

Schedule the Run

Scroll to **Build Triggers** and tick **Build periodically**.

In the cron box, enter the code shown in Code-block 11-13. This executes the job every day at **09:30 UTC**. Adjust the schedule to match your own window.

Code-block 11-13. Cron schedule syntax

```
30 9 * * *
```

Define the Build Steps

Under **Build**, click **Add build step ➤ Execute shell** and paste content from Code-block 11-14.

Code-block 11-14. Build Steps Shell commands

```
cd /app/etl_scripts/
python etl_job.py --source aws
```

The first line moves into the folder mapped by Docker; the second launches the ETL for AWS and lets the script decide which date to process; see Code-block 11-15. If you ever need a back-fill, you can add the `--date YYYY-MM-DD` flag, for example, see Code-block 11-15.

Code-block 11-15. Build Steps Shell commands

```
cd /app/etl_scripts/
python etl_job.py --source aws --date 2025-05-18
```

Save and Test

Click **Save**; see Figure 11-5. Jenkins shows the job page with a **Build Now** link in the sidebar. Use it to trigger a run immediately. Notice that Figure 11-5 is running with Code-block 11-14.

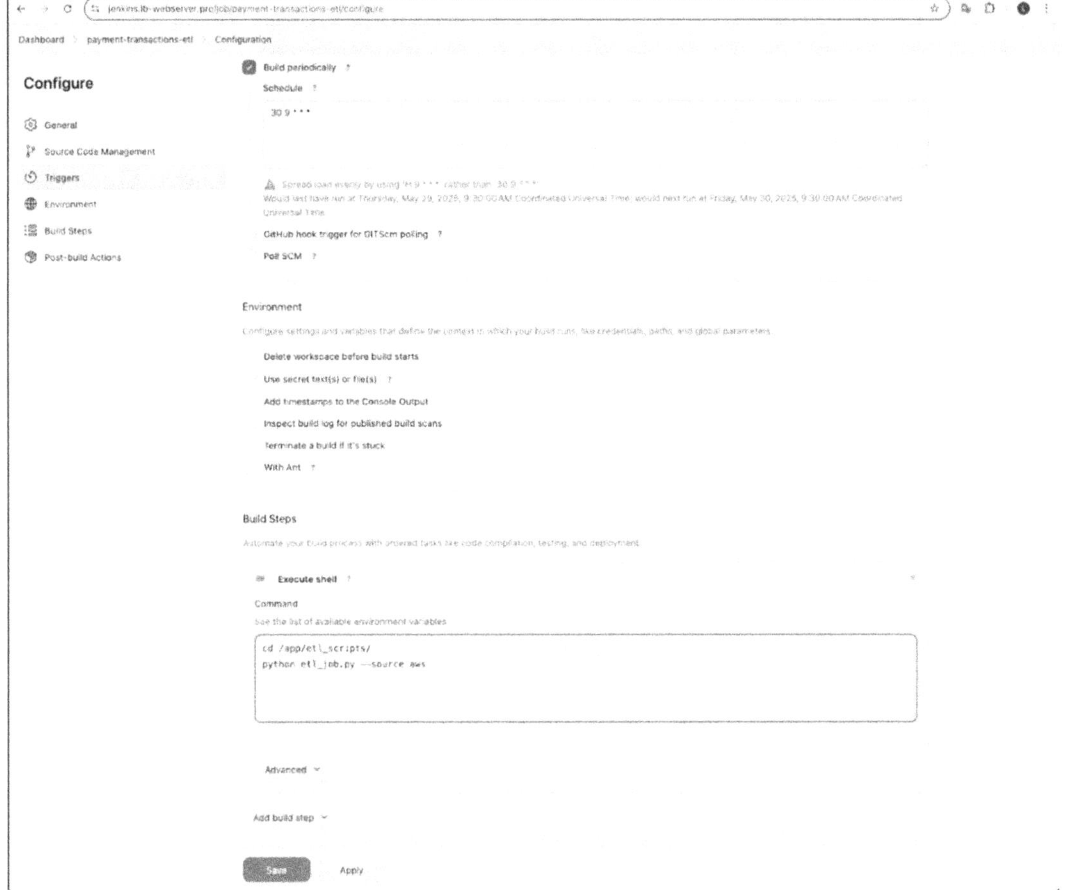

Figure 11-5. *Define the schedule and the shell commands.*

- If the raw files are present, the build will turn green.

- If any file is missing, the build fails; see Figure 11-6. You can inspect the logs to see the reason; in this case, it's a NoSuchKey error (see Code-block 11-16); the console log pinpoints the absent object.

Code-block 11-16. Build #1 console output

```
Started by user Lucas Braga

Running as SYSTEM
Building in workspace /var/jenkins_home/workspace/payment-transactions-etl
[payment-transactions-etl] $ /bin/sh -xe /tmp/
jenkins12284877394855064480.sh
+ cd /app/etl_scripts/
+ python etl_job.py --source aws
Extracting raw_transactions_2025_05_28.csv from AWS S3 bucket book-
project-54641
Traceback (most recent call last):
  File "/app/etl_scripts/etl_job.py", line 149, in <module>
    run_etl(source, today, CREDENTIALS_PATH)
  File "/app/etl_scripts/etl_job.py", line 110, in run_etl
    df_ls.append(extract_from_s3(AWS_BUCKET_NAME, file))
                 ^^^^^^^^^^^^^^^^^^^^^^^^^^^^^^^^^^^^^^^^
  File "/app/etl_scripts/etl_job.py", line 19, in extract_from_s3
    response = s3.get_object(Bucket=bucket_name, Key=file_name)
               ^^^^^^^^^^^^^^^^^^^^^^^^^^^^^^^^^^^^^^^^^^^^^^^^^^
  File "/opt/venv/lib/python3.11/site-packages/botocore/client.py", line
565, in _api_call
    return self._make_api_call(operation_name, kwargs)
           ^^^^^^^^^^^^^^^^^^^^^^^^^^^^^^^^^^^^^^^^^^^^^^
  File "/opt/venv/lib/python3.11/site-packages/botocore/client.py", line
1017, in _make_api_call
    raise error_class(parsed_response, operation_name)
botocore.errorfactory.NoSuchKey: An error occurred (NoSuchKey) when calling
the GetObject operation: The specified key does not exist.
Build step 'Execute shell' marked build as failure
Finished: FAILURE
```

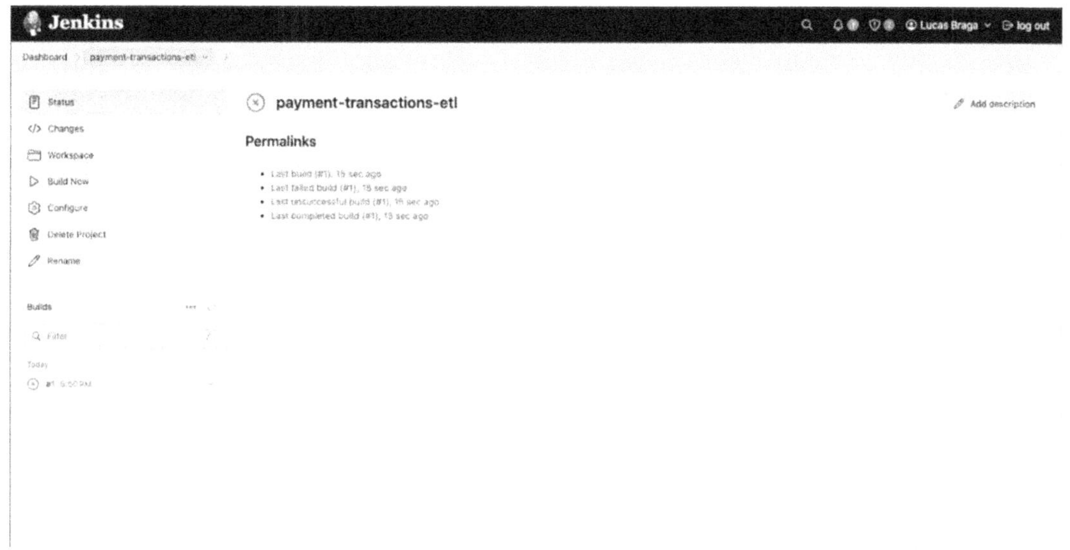

Figure 11-6. *Execute it manually*

You can edit the job at any time via **Configure**, handy when you want to run a one-off back-fill with a hard-coded `--date`. Adjust the shell command so that it uses the code from Code-block 11-15 instead. Then save it and click `Build Now` again to see that the Build #2 was successful; see Figure 11-7.

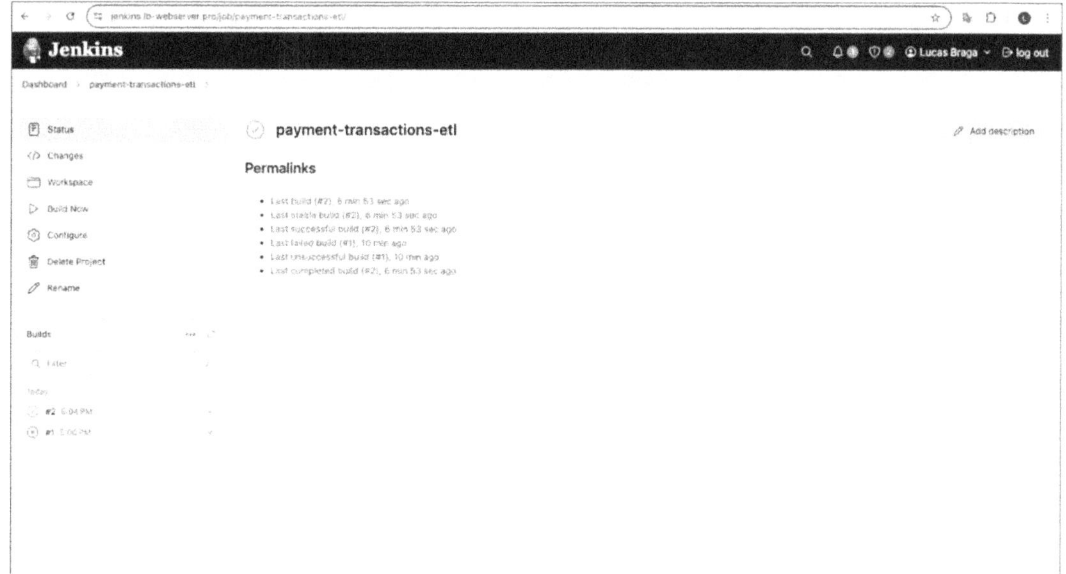

Figure 11-7. *Result after adjusting the configuration*

Verify the Output

After a successful build, open your storage console(s) and check for two new CSVs in the /output/ folder: `success_rate_df_YYYY_MM_DD.csv` and `amount_success_rate_df_YYYY_MM_DD.csv`; see Figures 11-8 and 11-9. Their presence confirms the pipeline extracted, transformed, and loaded the data exactly as designed.

Your ETL is now fully automated: Jenkins wakes up at 09:30 UTC, runs the script, and lands the processed metrics where analysts, or downstream dashboards, can pick them up.

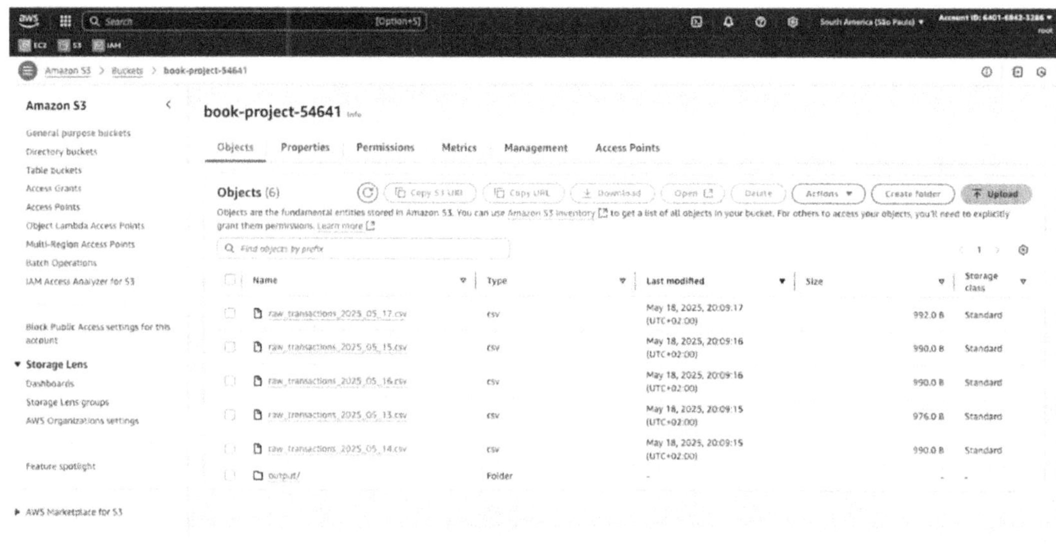

Figure 11-8. *AWS S3 bucket*

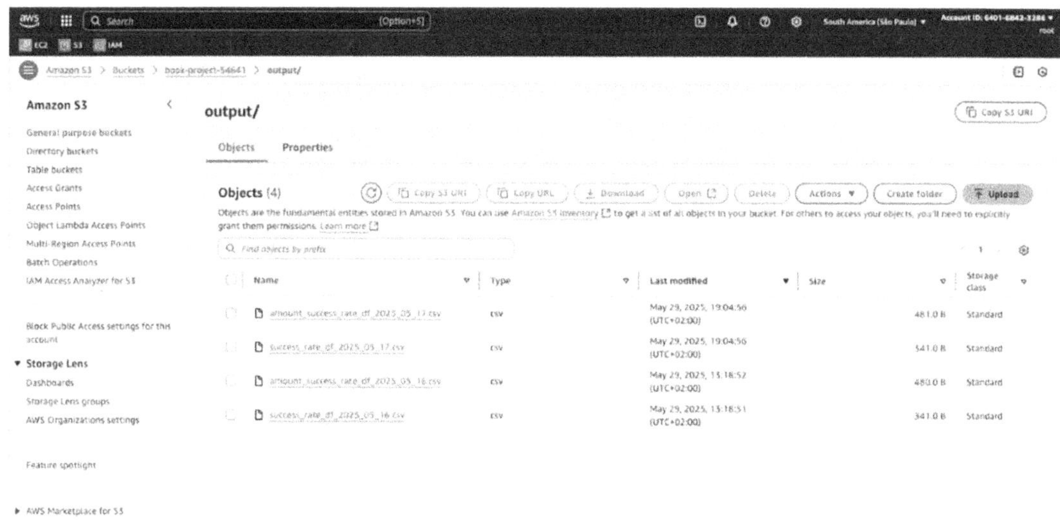

Figure 11-9. *AWS S3 bucket output folder files*

Summary

In this chapter, you implemented a hands-free ETL/ELT pipeline using Jenkins and Python. Specifically, you

- Understood the difference between **ETL** (Extract → Transform → Load) and **ELT** (Extract → Load → Transform) and when each pattern is appropriate

- Created AWS and GCP **credentials** that Jenkins could use to read and write cloud storage buckets

- Configured **S3** and **Google Cloud Storage** buckets and uploaded raw transaction files to simulate daily data ingestion

  - Boto3 AWS SDK documentation: `https://boto3.amazonaws.com/v1/documentation/api/latest/index.html`

  - Google Cloud Storage Python Client documentation: `https://cloud.google.com/python/docs/reference/storage/latest`

- Wrote a Python-based ETL pipeline (`etl_job.py`) that

  - Extracts CSV files from the correct cloud bucket

  - Cleans, deduplicates, transforms, and aggregates metrics

  - Uploads results back to cloud buckets as processed CSVs

- Created a Jenkins **Freestyle project** to

  - Bind mount credentials and ETL scripts inside the Jenkins container

  - Trigger daily builds using cron syntax

    - Cron syntax (Jenkins) documentation: `https://www.jenkins.io/doc/book/pipeline/syntax/#cron-syntax`

  - Execute the ETL pipeline with optional CLI parameters (e.g., --date for backfills)

    - Jenkins Freestyle Projects documentation: `https://www.jenkins.io/doc/book/pipeline/getting-started/`

- Validated results by confirming the output CSVs landed correctly in `/output/` folders on AWS S3 and GCP buckets

In the next chapter, you'll explore how to use **Streamlit** as a lightweight, interactive **dashboarding tool** to visualize the data products your ETL pipeline generates. Following that, Chapter 13 will show how to use **Flask** to serve a trained ML model through a real-time prediction API, completing the end-to-end deployment of your data science stack.

## CHAPTER 12

Demo: Streamlit

In the previous chapter, you built an automated ETL pipeline using Jenkins and Python. The job pulled raw transaction data from cloud storage, cleaned and aggregated it, and saved daily KPIs into new output folders on both AWS S3 and Google Cloud Storage.

In this chapter, you'll build a lightweight dashboard using Streamlit that reads those cloud buckets and visualizes success rates and revenue metrics over time. The dashboard will allow stakeholders to filter by date, payment service provider (PSP), and metric type, all from a simple, code-based interface. By the end, you'll have a live dashboard that updates automatically as soon as new ETL results are available, completing the end-to-end flow from data ingestion to user-facing visualization.

Objective and Setup Overview

Streamlit has become a very popular tool among data scientists because it turns pure Python scripts into interactive web apps with almost no overhead. The same code base can serve as dashboard, a lightweight website, or even a front end for machine-learning models that return real-time predictions. This code-first approach offers three major advantages:

- **Familiar tooling:** Build everything in Python, no need to learn HTML, JavaScript, or traditional BI platforms.

- **Reproducibility:** Version-controlled apps that run identically across environments with "streamlit run".

- **Developer-friendliness:** Works seamlessly with pandas, NumPy, and cloud SDKs.

© Lucas H. Benevides e Braga 2025
L.H.B.e. Braga, *Deploying Secure Data Science Applications in the Cloud,*
https://doi.org/10.1007/979-8-8688-1715-1_12

In this chapter, you will stay within the same fintech scenario introduced in Chapter 11. The Jenkins ETL job creates daily output files in S3 and GCP. Your new Streamlit dashboard will read those files and let users drill into specific PSPs and dates, and surface metrics with tables and charts. Figure 12-1 shows the finished app you'll build by the end of the chapter.

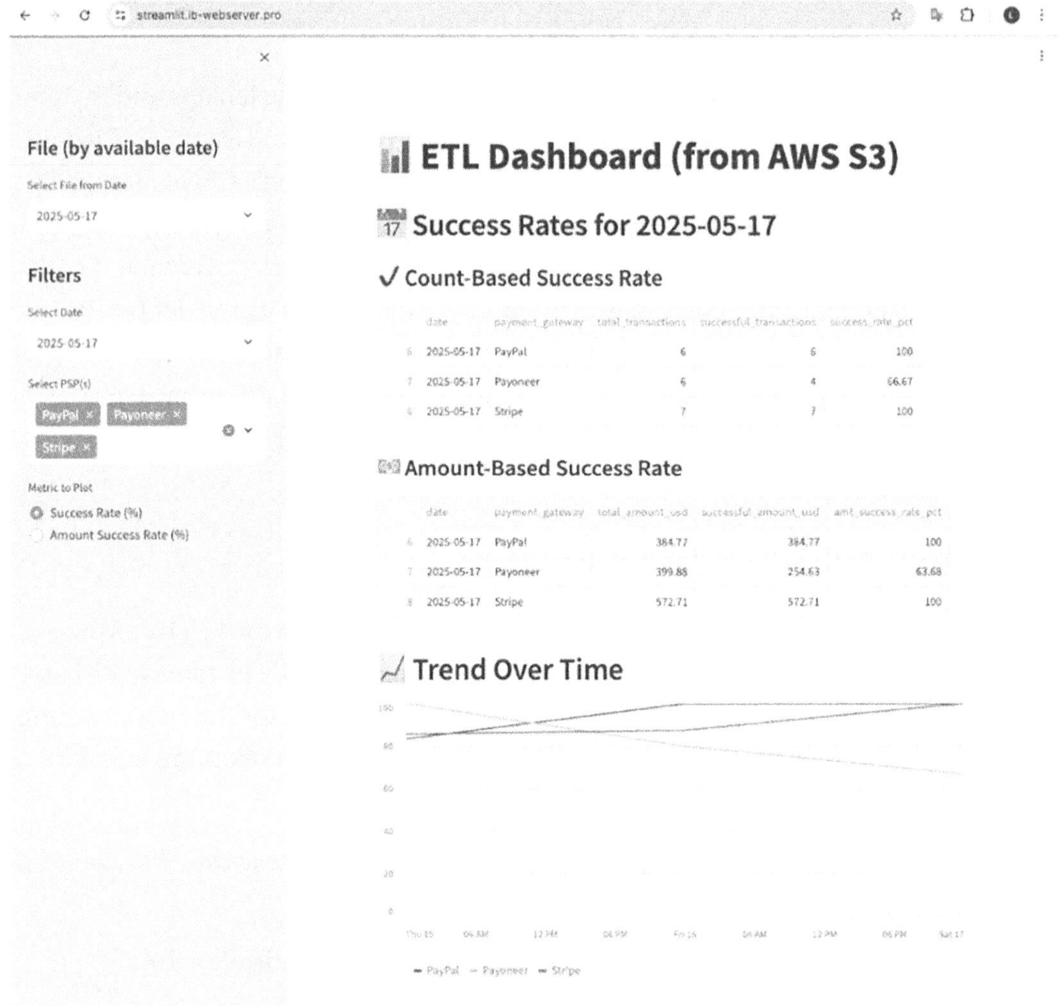

Figure 12-1. *Streamlit Dashboard end result*

Docker and Infrastructure Configuration

Before we delve into the Streamlit dashboard code, you'll need to update your existing web-server stack to replace the old calculator app with the new **streamlit_dashboard** container. This section walks through four required updates: Docker Compose, Nginx routing, dependencies, and the app's Dockerfile. **Update docker-compose.yaml.**

Start with **docker-compose.yaml**. The stanza that used to build *streamlit_calc* is now commented out, and a new service called **streamlit_dashboard** takes its place. Everything else remains unchanged: the container still publishes port 8501, restarts automatically, and sits behind Nginx. The crucial addition is the environment variable AWS_CONFIG_FILE=/app/.aws/config, which exposes the AWS profile you copied into the dashboard directory during the credential-setup step.

Also ensure you have placed the GCP service-account key under **streamlit_dashboard/config/** so the app can talk to Google Cloud Storage; see Code-block 12-1.

Code-block 12-1. Updated docker-compose.yaml

```yaml
version: "3.7"

services:
 # streamlit_calc:        # old app — now retired
 # ...

 streamlit_dashboard:
   build:
     context: streamlit_dashboard
   ports:
     - "8501"
   environment:
     - AWS_CONFIG_FILE=/app/.aws/config
   restart: always

# ...

 Jenkins:
 # ...
     # - ./streamlit_calc/data:/app/data
 # ...
```

```
nginx:
# ...
   depends_on:
     - flask
    # - streamlit_calc
     - streamlit_dashboard
     - jenkins
# ...
```

Update nginx.conf

Because the service name changed, you must update **nginx.conf** as well. Only one line moves: proxy_pass now points to streamlit_dashboard instead of streamlit_calc; see Code-block 12-2.

Code-block 12-2. nginx.conf changes one line

```
# proxy_pass http://streamlit_calc:8501;
proxy_pass http://streamlit_dashboard:8501;
```

Add Python Dependencies

Next, list the dashboard's Python dependencies in **streamlit_dashboard/ requirements.txt**. They mirror the ETL job except for the addition of streamlit and a newer Tornado version, which Streamlit requires. See Code-block 12-3.

Code-block 12-3. Requirements.txt for streamlit_dashboard

```
streamlit==1.34.0
tornado==6.4
pandas==2.0.3
numpy==1.24.4
boto3==1.20.0
google-cloud-storage==2.0.0
```

Modify the Dockerfile

Finally, tweak the **Dockerfile** in the same folder so the last line launches *app.py* instead of the calculator script; see Code-block 12-4.

Code-block 12-4. Dockerfile from streamlit_dashboard one line changes

```
CMD ["streamlit", "run", "--server.address", "0.0.0.0", "--server.port",
"8501", "--server.enableCORS", "false", "--server.enableXsrfProtection",
"true", "app.py"]
```

With those four changes—compose file, Nginx route, requirements list, and Dockerfile—you can rebuild the stack (`docker-compose up --build -d`). When the containers settle, browsing to `https://streamlit.lb-webserver.pro` will load the new dashboard, ready to consume the metrics you generated in Chapter 11.

App Architecture and Code Walkthrough

The dashboard logic lives in **app.py**. Much like the ETL script from Chapter 11, it begins by pulling in the cloud SDKs and defining a handful of constants, this time pointing to the **output/** folder that the ETL job populates each morning. See Code-block 12-5.

Code-block 12-5. app.py initial part

```python
import streamlit as st
import pandas as pd
from io import BytesIO
from pathlib import Path
import boto3
from google.cloud import storage

# === CONFIG ===
AWS_BUCKET_NAME = 'book-project-54641'
GCP_BUCKET_NAME = 'book-etl-project'
GCP_CREDENTIALS_PATH = Path("config/lb-cloud-project-65885ec00848.json")
DATA_FOLDER = "output"
SOURCE = 'aws' # Change to 'gcp' for Google Cloud Storage or to 'aws'
for AWS S3
```

Breaking down these constants:

- **AWS_BUCKET_NAME** and **GCP_BUCKET_NAME** match the buckets you created in Chapter 11, section "Create Buckets and Upload Test Files."

- **GCP_CREDENTIALS_PATH** locates the .json credentials to access the GCP bucket.

- **DATA_FOLDER** is fixed at **output/**, the subdirectory where etl_job.py writes its results.

- **SOURCE** flips between **aws** and **gcp**, letting you test the dashboard against either cloud without touching the code further down.

Code-block Files and Available Dates

From here, the script will

1. Connect to the chosen bucket (using boto3 or google.cloud.storage)

2. Read the two KPI CSVs: success_rate_df_... and amount_success_rate_df_... into pandas

3. Render tables, charts, and filters with Streamlit widgets

We'll unpack those steps in the sections that follow, but the headline is straightforward: the dashboard talks to the same buckets the ETL job writes to, reads yesterday's fresh metrics, and turns them into an interactive view for stakeholders.

The dashboard needs three building blocks before it can draw any charts: a way to **discover** which CSVs exist, a way to **parse** the date embedded in each file name, and finally a way to **download** the chosen files into pandas. Those tasks are handled by the helpers in Code-blocks 12-6 and 12-7.

Code-block 12-6. app.py functions to list file names

```
# === Utility: List S3 files ===
def list_files_s3(bucket_name, prefix):
    s3 = boto3.Session(profile_name='book-etl').client('s3')
    response = s3.list_objects_v2(Bucket=bucket_name, Prefix=prefix)
    files = [obj['Key'] for obj in response.get('Contents', [])]
    return files
```

```python
# === Utility: List GCP files ===
def list_files_gcp(bucket_name, prefix, credentials_path):
    client = storage.Client.from_service_account_json(str(credentials_path))
    bucket = client.get_bucket(bucket_name)
    blobs = bucket.list_blobs(prefix=prefix)
    return [blob.name for blob in blobs]

# === Extract dates from files available ===
def extract_dates_from_filenames(file_list, pattern=r"_(\d{4}_\
d{2}_\d{2})\.csv$"):
    # Extract the last 10 characters before '.csv'
    dates = []
    for filename in file_list:
        m = re.search(pattern, filename)
        if m:
            dates.append(m.group(1))
    return sorted(list(set(dates)), reverse=True)
```

- `list_files_s3()` and `list_files_gcp()` return every object key that begins with output/, which is where the ETL job writes its results.

- `extract_dates_from_filenames()` isolates the YYYY_MM_DD portion so the UI can present a tidy date picker.

Loading the Data from Cloud Storage

Once the user selects a date, the dashboard needs to fetch two specific files, `success_rate_df_*.csv` and `amount_success_rate_df_*.csv`, and turn them into DataFrames. That work is done in the next trio of helpers; see Code-block 12-7.

Code-block 12-7. app.py functions to extract file from buckets

```python
# === Utility: Load from S3 ===
def load_csv_from_s3(bucket_name, file_key):
    s3 = boto3.Session(profile_name='book-etl').client('s3')
    response = s3.get_object(Bucket=bucket_name, Key=file_key)
    return pd.read_csv(BytesIO(response['Body'].read()), parse_
    dates=["date"])
```

```python
# === Utility: Load from GCP ===
def load_csv_from_gcp(bucket_name, file_key, credentials_path):
    client = storage.Client.from_service_account_json(str(credentials_path))
    bucket = client.get_bucket(bucket_name)
    blob = bucket.blob(file_key)
    return pd.read_csv(BytesIO(blob.download_as_bytes()),
    parse_dates=["date"])

# === Load data from cloud ===
@st.cache_data
def load_data(source, date_str):
    success_file = f"{DATA_FOLDER}/success_rate_df_{date_str}.csv"
    amount_file = f"{DATA_FOLDER}/amount_success_rate_df_{date_str}.csv"

    if source == "aws":
        success_df = load_csv_from_s3(AWS_BUCKET_NAME, success_file)
        amount_df = load_csv_from_s3(AWS_BUCKET_NAME, amount_file)
    else:
        success_df = load_csv_from_gcp(GCP_BUCKET_NAME, success_file,
        GCP_CREDENTIALS_PATH)
        amount_df = load_csv_from_gcp(GCP_BUCKET_NAME, amount_file,
        GCP_CREDENTIALS_PATH)

    return success_df, amount_df
```

st.cache_data tells Streamlit to store the returned DataFrames in memory for
the life of the session, so navigating through different widgets won't trigger redundant
network calls.

With these utilities in place, the remainder of app.py can focus on layout: building
date selectors, drop-downs for payment-gateway filters, and the charts themselves,
confident that the raw numbers will arrive with one function call.

Interactive Dashboard Logic

The remainder of **app.py** turns your utility functions into an interactive dashboard.
Because every Streamlit command is prefixed with st., you can read the code almost
like plain English.

File Section Sidebar

The first call adds a header to the sidebar; see Code-block 12-8. Immediately after, the script queries the bucket for every success-rate file so it can offer a list of available snapshots.

all_files is just a list of keys, so the helper extract_dates_from_filenames() pulls out the YYYY_MM_DD segment and sorts it. Each string then becomes a datetime.date object to make the drop-down more readable.

The moment a user chooses a date, load_data() fetches the two KPI CSVs for that snapshot and caches them.

A second st.sidebar.title() introduces the dynamic filters. The code merges the unique dates from both DataFrames, sorts them in reverse order, and offers a **Date** selector. Because each ETL output aggregates three days' worth of raw data, this lets a user drill into a particular trading day without loading a new file.

Code-block 12-8. app.py file section sidebar

```
# === Sidebar Filters ===
st.sidebar.title("File (by available date)")
# List available dates in the bucket
if SOURCE == 'aws':
    all_files = list_files_s3(AWS_BUCKET_NAME, f"{DATA_FOLDER}/success_
    rate_df_")
else:
    all_files = list_files_gcp(GCP_BUCKET_NAME, f"{DATA_FOLDER}/success_
    rate_df_", GCP_CREDENTIALS_PATH)

all_dates_str = extract_dates_from_filenames(all_files)
# Convert to datetime.date objects for display/selection
all_file_dates = [pd.to_datetime(date, format="%Y_%m_%d").date() for date
in all_dates_str]

selected_file_from_date = st.sidebar.selectbox("Select File from Date",
options=all_file_dates)
selected_date_str = selected_file_from_date.strftime("%Y_%m_%d")

success_df, amount_df = load_data(SOURCE, selected_date_str)
```

```
st.sidebar.title("Filters")

all_dates = pd.concat([success_df["date"], amount_df["date"]]).drop_
duplicates().sort_values(ascending=False)
selected_date = st.sidebar.selectbox("Select Date", options=all_dates.
dt.date)
```

Filter by Date and PSP

Next comes the **Payment Service Provider (PSP)** filter, as shown in Code-block 12-9. The code unions the payment_gateway columns from both metrics, sorts them, and feeds the list to st.sidebar.multiselect(). By default, every PSP is selected.

Finally, a radio button lets the viewer choose which metric to plot.

At runtime the sidebar looks like Figure 12-2. First you pick the **file date**, for example, *2025-05-17*. Behind the scenes, the dashboard loads.

Figure 12-3 shows the remaining filters: **Date**, **PSP(s)**, and **Metric**. Adjusting any of these widgets reruns the downstream Streamlit code and refreshes the tables and charts in the main panel, giving stakeholders an instant, self-service view of payment performance.

Code-block 12-9. app.py filter by date and PSP

```
all_psps = sorted(set(success_df["payment_gateway"]) |
set(amount_df["payment_gateway"]))
selected_psps = st.sidebar.multiselect("Select PSP(s)", options=all_psps,
default=all_psps)

metric = st.sidebar.radio("Metric to Plot", options=["Success Rate (%)",
"Amount Success Rate (%)"])
```

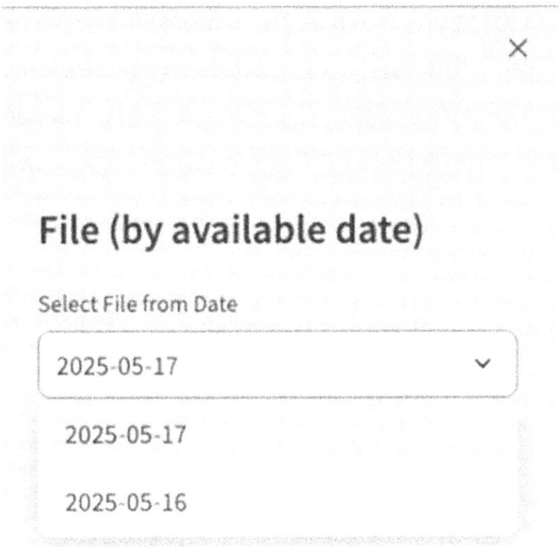

Figure 12-2. *See menu to choose file to extract*

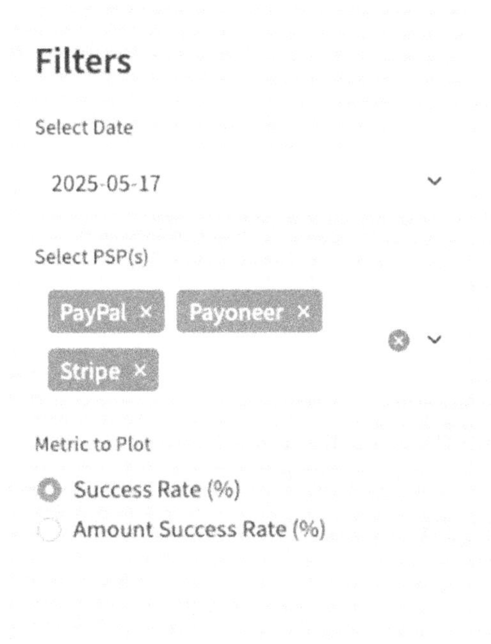

Figure 12-3. *Sidebar filters*

Selecting a new date in the sidebar immediately refreshes everything on-screen. When a reader chooses **2025-05-16**, Streamlit stores that choice in selected_date, and then the code in Code-block 12-9 rebuilds the tables on the fly. Two Boolean masks pick only the rows whose date equals that value and whose payment_gateway appears in selected_psps. The script then converts the timestamp to a neat YYYY-MM-DD string, renames a couple of verbose columns, and renders the result with st.dataframe(). Because Streamlit reruns the script top to bottom on every interaction, the new data arrive in milliseconds, and Figure 12-4 updates without a page reload.

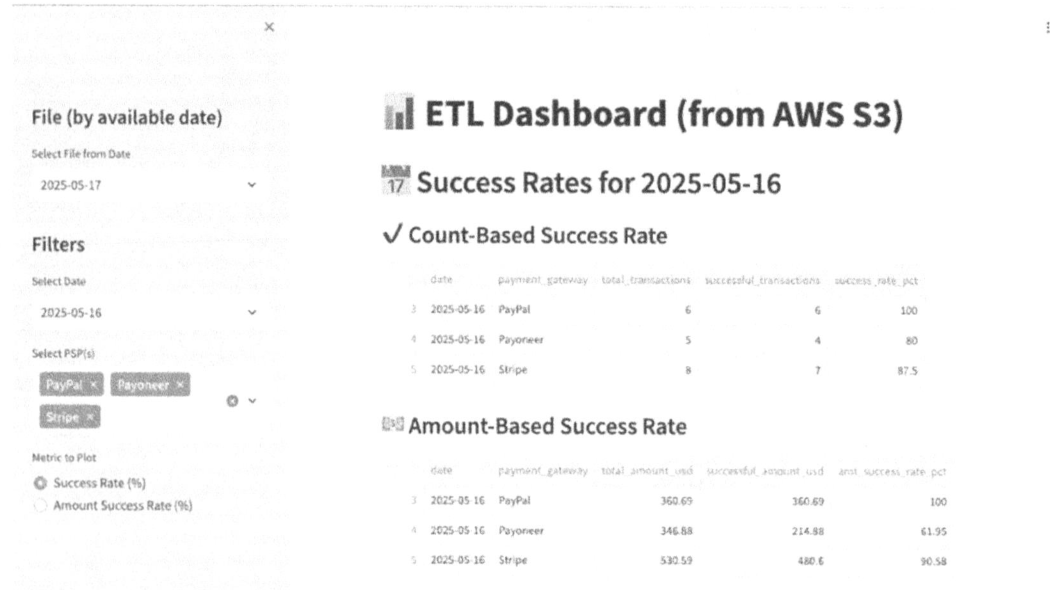

Figure 12-4. *Main tables with date filtered to 2025-05-16*

Table Rendering

After filtering the data by selected date and PSPs, the dashboard displays two separate tables: one for count-based success rates and another for amount-based success rates. The date column is formatted for readability, and the metric columns are renamed to cleaner labels. These tables give users a snapshot of payment performance for a specific day. See Code-block 12-10.

Code-block 12-10. How sidebar filters interactively change the tables

```python
# === Filter data ===
filtered_success_df = success_df[
    (success_df["date"].dt.date == selected_date) &
    (success_df["payment_gateway"].isin(selected_psps))
]
```

```
filtered_amount_df = amount_df[
    (amount_df["date"].dt.date == selected_date) &
    (amount_df["payment_gateway"].isin(selected_psps))
]

# === Display Tables ===
if SOURCE == "aws":
    st.title("📊 ETL Dashboard (from AWS S3)")
else:
    st.title("📊 ETL Dashboard (from Google Cloud Storage)")

st.header(f"📅 Success Rates for {selected_date}")

# Format date for display in tables
filtered_success_df["date"] = filtered_success_df["date"].
dt.strftime('%Y-%m-%d')
filtered_amount_df["date"] = filtered_amount_df["date"].
dt.strftime('%Y-%m-%d')

filtered_success_df.rename(columns={
    "success_rate_percent": "success_rate_pct",
}, inplace=True)
filtered_amount_df.rename(columns={
    "amount_success_rate_percent": "amt_success_rate_pct"
}, inplace=True)

st.subheader("✓ Count-Based Success Rate")
st.dataframe(filtered_success_df.sort_values("payment_gateway"))

st.subheader("💰 Amount-Based Success Rate")
st.dataframe(filtered_amount_df.sort_values("payment_gateway"))
```

Line Chart Over Time

The final element is a time-series plot that tracks performance by provider. First, the code merges the count-based and amount-based DataFrames so both metrics live in one table. It filters the merge to only the gateways the user ticked in the multiselect, pivots the table so each column represents a PSP, and passes the result to st.line_chart(); see Code-block 12-11.

Depending on whether the *Metric* radio button is set to **Success Rate (%)** or **Amount Success Rate (%)**, the chart traces the appropriate series. Figure 12-5 shows a typical view with PayPal, Payoneer, and Stripe toggled on.

Code-block 12-11. app.py line graphs at the bottom

```python
# === Line Chart ===
st.header("📈 Trend Over Time")

merged_df = pd.merge(
    success_df[["date", "payment_gateway", "success_rate_percent"]],
    amount_df[["date", "payment_gateway", "amount_success_rate_percent"]],
    on=["date", "payment_gateway"],
    how="outer"
)

merged_filtered = merged_df[merged_df["payment_gateway"].
isin(selected_psps)]

if metric == "Success Rate (%)":
    st.line_chart(
        merged_filtered.pivot_table(index="date", columns="payment_gateway",
        values="success_rate_percent")
    )
else:
    st.line_chart(
        merged_filtered.pivot_table(index="date", columns="payment_gateway",
        values="amount_success_rate_percent")
    )
```

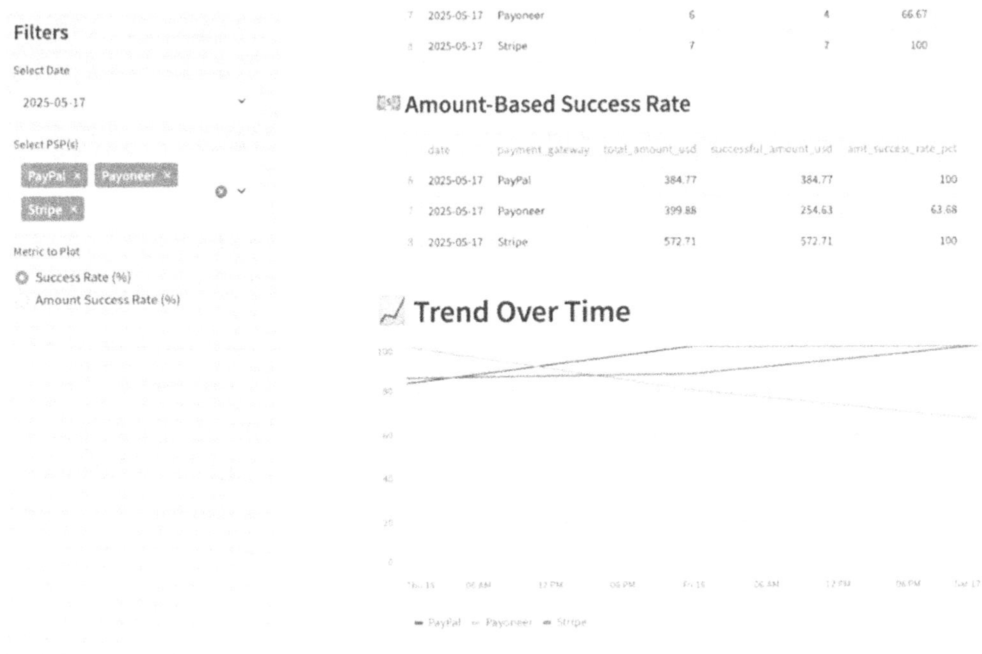

Figure 12-5. *Line graph at the bottom of the Dashboard*

Summary

In this chapter, you created a reproducible, cloud-connected dashboard using Streamlit. The dashboard read from the same S3 and GCP buckets populated by your Jenkins ETL job, parsed daily KPI files, and let users explore results with filters and visualizations. You updated your Docker setup, Nginx proxy, and Streamlit service definition, deploying the new app behind an HTTPS-secured subdomain.

We walked through the configuration changes, new service stanza in *docker-compose.yaml*, a one-line tweak to *nginx.conf*, an updated *requirements.txt*, and a Dockerfile that points to *app.py*. Inside the Python script we explored bucket discovery, secure file download, on-the-fly filtering, dataframe display, and real-time charting, all driven by Streamlit's intuitive `st.` package.

Reference documentation:

- Streamlit documentation: `https://docs.streamlit.io/`

- Google Cloud Storage Python Client documentation: `https://cloud.google.com/python/docs/reference/storage/latest`

- Boto3 AWS SDK documentation for S3: `https://boto3.amazonaws.com/v1/documentation/api/latest/reference/services/s3.html`

The result is a fully reproducible, code-first dashboard that refreshes as soon as the ETL job drops fresh data into the bucket, giving your stakeholders an up-to-date view of payment performance every morning. In the next chapter, you'll build a Flask-based prediction API. We'll examine the container layout and how Gunicorn manages request handling for Flask and walk through the application code that serves machine learning predictions to external clients.

## CHAPTER 13

Demo: Flask

Back in Chapter 5 you trained a classifier, serialized the model object, and invoked it from a notebook. In this chapter, you will step inside the containerized Flask service that now hosts that model. We will examine how the code, Docker image, and infrastructure layers fit together so you can adapt the pattern for your own projects.

Flask gives data scientists a lightweight path from Python code to a production-ready HTTP API. Package your trained model inside a Flask app and expose a /predict endpoint, and any client–mobile, web, or another service–can request live predictions at scale.

Container Layout

Before diving into the code, take a minute to orient yourself inside the **flask/** directory, see Code-block 13-1. Every file serves a specific role in turning a pickled model into a production-grade API.

Code-block 13-1. Folder structure

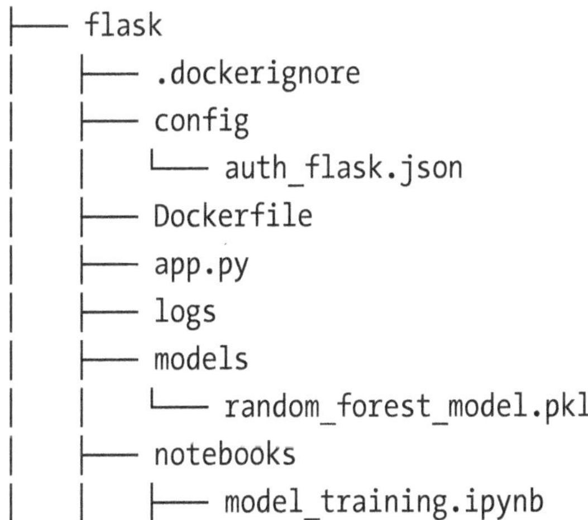

```
├── flask
│   ├── .dockerignore
│   ├── config
│   │   └── auth_flask.json
│   ├── Dockerfile
│   ├── app.py
│   ├── logs
│   ├── models
│   │   └── random_forest_model.pkl
│   ├── notebooks
│   │   ├── model_training.ipynb
```


© Lucas H. Benevides e Braga 2025
L.H.B.e. Braga, *Deploying Secure Data Science Applications in the Cloud,*
https://doi.org/10.1007/979-8-8688-1715-1_13

```
|    |    └── post_predict.ipynb
|    ├── requirements.txt
|    ├── templates
|    |    ├── index-without-subdomains.html
|    |    └── index.html
|    └── wsgi.py
```

- **app.py** – The heart of the service. It loads the pickled **random_forest_model.pkl**, parses the credentials in **config/auth_flask.json**, and defines HTTP endpoints (/ and /predict) plus their allowed methods (GET and POST).

- **models/random_forest_model.pkl** – The trained Random Forest you exported in Chapter 5. Flask deserializes this object on startup so predictions are instantaneous.

- **Dockerfile** – Declares the runtime image, installs the dependencies in **requirements.txt**, and launches **Gunicorn** rather than Flask's built-in server. Gunicorn is a production-grade WSGI server that can spawn multiple worker processes, handle thousands of concurrent requests, and keep your API alive under load.

- **wsgi.py** – A tiny shim that exposes the Flask application instance. Gunicorn reads this file (not *app.py* directly) when it starts, giving you the clean separation encouraged by the WSGI spec.

- **config/auth_flask.json** – Holds whatever API credentials the service needs—keys, tokens, or static user/password pairs. Keeping secrets in a separate file makes it easy to swap them out for different environments.

- **templates/** – Simple Jinja2 HTML files for the landing page. They are optional for a pure API but handy if you want a "Hello, world—service is alive" front page.

- **notebooks/** – The original exploratory work: one notebook that trains the model and another that shows how to call the /predict endpoint once the container is running.

When Docker builds the image, it adds everything under **flask/**, installs the listed libraries, and then launches

- `gunicorn --bind 0.0.0.0:8502 wsgi:app`

Gunicorn dutifully imports app from `wsgi.py` and spins up its workers, and your model becomes an HTTP service ready to answer real-time prediction requests.

How Gunicorn Hands Requests to Flask

Think of Gunicorn as the production-grade engine that sits between incoming HTTP requests and your Flask code, much the way Nginx buffers traffic for static content. At container startup, the last line of the **Dockerfile** launches Gunicorn and tells it exactly where to find the Flask application; see Code-block 13-2.

Code-block 13-2. Last line of the Dockerfile

```
CMD ["gunicorn", "--bind", "0.0.0.0:8502", "wsgi:app", "timeout", "120",
"--workers=3"]
```

- `--bind 0.0.0.0:8502` opens the service on port 8502 inside the container.

- `--workers 3` instructs Gunicorn to fork three worker processes, giving you parallel request handling out of the gate.

- `--timeout 120` sets a two-minute ceiling on any single request, long enough for most ML inference yet short enough to prevent worker lock-ups.

- `wsgi:app` is the critical pointer: **wsgi** is the module name (`wsgi.py`) and **app** is the Flask application object inside that module.

The companion file **wsgi.py** couldn't be simpler; see Code-block 13-3.

Code-block 13-3. wsgi.py file

```
from app import app

if __name__=='__main__':
    app.run()
```

When Gunicorn evaluates `wsgi:app`, it imports `wsgi.py`, which in turn imports the app object from **app.py**. Gunicorn then keeps that Flask instance alive and delegates each HTTP request to whichever worker is free.

Two conventions make the hand-off seamless:

1. **File naming:** Your Flask script is called **app.py**, so `from app import app` does exactly what it says.

2. **WSGI entry point:** The single line `wsgi:app` in the Gunicorn command points to that same object, no extra wiring required.

With those pieces in place, Gunicorn handles concurrency, worker restarts, and graceful shutdown, while Flask focuses solely on parsing JSON bodies, calling the model, and returning predictions.

Flask Service Walkthrough

The API is only a few dozen lines, but each line does real work—from pulling the model into memory to authenticating callers and returning a prediction.

Loading the Assets

`app.py` opens with the usual imports plus two essentials: **pickle** to deserialize the Random Forest model and **json** to parse the credential file stored in `config/auth_flask.json`. The code then tries to load both artifacts, raising a helpful log message if either one is missing; see Code-block 13-4. At the end of this section, two global variables exist:

- **model_obj** – The ready-to-score estimator

- **users** – A dictionary that holds the production username and password

Code-block 13-4. app.py initial part

```
from flask import Flask, render_template, request, jsonify
from flask_httpauth import HTTPBasicAuth
import pickle
```

```python
import numpy as np
import json
import logging

# Load model and load Auth keys
HOME_PATH = '.'
PATH_MODEL = f'{HOME_PATH}/models/random_forest_model.pkl'
PATH_AUTH = f'{HOME_PATH}/config/auth_flask.json'
try:
    with open(PATH_MODEL, 'rb') as file:
        model_obj = pickle.load(file)
except FileNotFoundError:
    logging.error(f"Model file not found at {PATH_MODEL}")
    raise
except Exception as e:
    logging.error(f"Error loading model: {str(e)}")
    raise

try:
    with open(PATH_AUTH, 'r') as file:
        users = json.load(file)
except FileNotFoundError:
    logging.error(f"Auth file not found at {PATH_AUTH}")
    raise
except Exception as e:
    logging.error(f"Error loading auth file: {str(e)}")
    raise
```

Creating the Flask App and the Auth Layer

Next, a Flask instance called **app** and an **HTTPBasicAuth** helper called **auth** come to life. Logging is configured to drop everything–headers, payloads, errors–into logs/flask. log, giving you a breadcrumb trail when something goes wrong; see Code-block 13-5.

The verify_password() callback cross-checks incoming credentials against the values loaded from auth_flask.json. Successful log-ins return the username; failures hit the warning log.

Code-block 13-5. app.py defines logging and authentication functions

```python
app = Flask('Score transactions')
auth = HTTPBasicAuth()

# Setup logging
logging.basicConfig(filename='logs/flask.log', level=logging.DEBUG,
format='%(asctime)s %(message)s')

@app.before_request
def log_request_info():
    logging.debug(f"Headers: {request.headers}")
    logging.debug(f"Body: {request.get_data()}")

# Setup password authentication
@auth.verify_password
def verify_password(username, password):
    logging.debug(f"Auth attempt with username={username} and
    password={password}")
    if username == users['prod']['user'] and password == users['prod']
    ['password']:
        logging.debug("Authentication successful")
        return username
    logging.warning(f"Failed auth attempt with username={username}")
```

Exposing the Endpoints

Two routes finish the service; see Code-block 13-6.

- **_/predict_** - _POST only, auth required_

  The body of the request is expected to be JSON—one key-value pair per feature. The handler converts that dictionary into a one-row NumPy array, feeds it to model_obj.predict_proba(), and returns the winning class's probability. Any exception (missing JSON, shape error, model glitch) yields a 400 response with an error message.

- **_/_** - _GET_

  A simple landing page rendered from templates/index.html; handy for uptime checks and quick demos.

316

Code-block 13-6. app.py final part

```python
# Setup predict function
@app.route('/predict', methods=['POST'])
@auth.login_required
def predict():
    try:
        data_input = request.get_json()
        if not data_input:
            raise ValueError("No input data provided")
        data_np = np.array(list(data_input.values())).reshape(1, -1)
        y_proba = model_obj.predict_proba(data_np)
        predicted_class = np.argmax(y_proba[0])
        predicted_probability = y_proba[0][predicted_class]
        result = {
            'model_name': 'random_forest_model',
            'score': predicted_probability
        }
        logging.debug(f"Prediction result: {result}")
        return jsonify(result), 200
    except Exception as e:
        logging.error(f"Error during prediction: {str(e)}")
        return jsonify({"error": str(e)}), 400

@app.route('/', methods=['GET'])
def home():
    return render_template(f"index.html")

#cd deploy-secure-ds-apps-book/flask && python3.9 app.py
if __name__ == "__main__":
    app.run(host="0.0.0.0", port=8502, debug=True)
```

When you run the service locally (python app.py), Flask binds to port 8502. Inside the container, however, **Gunicorn** launches the app via wsgi.py, spawns three workers, and enforces a 120-second timeout; see again Code-block 13-2. That swap gives you production-grade concurrency without changing a line of Flask code.

Summary

Over this chapter you assembled every moving part of a real-time inference API using Flask. You loaded a trained Random Forest model from Chapter 5, authenticated incoming requests, and returned live predictions through a **/predict** endpoint. You also implemented a basic landing page at **/** for simple uptime checks. Specifically, you

- Reviewed the structure of the **flask/** directory, including the model, config, templates, and notebooks

  - Flask documentation: `https://flask.palletsprojects.com/`

- Created a Flask app that handles GET and POST requests using **Flask** and **flask_httpauth**

  - Flask-HTTPAuth documentation: `https://flask-httpauth.readthedocs.io/en/latest/`

- Configured **Gunicorn** as the production-grade WSGI server for handling concurrent traffic

  - Gunicorn documentation: `https://docs.gunicorn.org/en/stable/`

- Implemented secure credential loading from a separate JSON file

- Set up logging for requests and responses via **logging** and Flask's **before_request** hook

- Exposed a **/predict** endpoint that accepts JSON payloads and returns a classification probability

- Defined a **wsgi.py** shim that cleanly connects Flask and Gunicorn, enabling scalable container deployment

  - WSGI documentation: `https://wsgi.readthedocs.io/en/latest/`

The result is a secure, scalable inference service you can run behind a load balancer or as part of a microservices architecture.

Index

A

I

J, K

L

M

N, O

P

GPSR Compliance
The European Union's (EU) General Product Safety Regulation (GPSR) is a set
of rules that requires consumer products to be safe and our obligations to
ensure this.

If you have any concerns about our products, you can contact us on

ProductSafety@springernature.com

In case Publisher is established outside the EU, the EU authorized
representative is:

Springer Nature Customer Service Center GmbH
Europaplatz 3
69115 Heidelberg, Germany